RHS

DK

SALLY NEX

How to
GARDEN
the
LOW
CARBON
way

THE STEPS YOU CAN TAKE TO HELP COMBAT CLIMATE CHANGE

CONTENTS

INTRODUCTION

Imagine a habitat of extraordinary diversity – capable of holding more plant varieties per square metre than a rainforest, as well as all kinds of wildlife. Sounds pretty amazing, doesn't it? And, what's more, it's right outside your back door.

Every one of our gardens is a microcosm of natural habitats in concentrated form. Yet they're rarely considered for their potential input to the wider environment, overlooked in the rush to study more romantic woodlands or meadows.

Least studied of all, it appears, is the role gardens may play in regulating our climate, and how gardeners might potentially be able to help them. Each and every garden is a tiny carbon sink, with every plant busily vacuuming carbon dioxide from the air then locking it away for years.

Nobody's ever worked out how much carbon our gardens are capable of absorbing each year. We do know how much they hold already, though: one estimate puts the amount of carbon tucked away in the top metre of garden soil at up to 145 tonnes (160 tons) of carbon per hectare. So the UK's million acres of gardens alone already lock away over 62 million tonnes (68 million tons) of carbon, and with our help they can sequester (capture and store) more.

As gardeners, we have a chance to make a direct, practical difference in combatting climate change. Every time you grow a tree, mulch your soil, or let your grass grow long, you could actively be increasing the carbon your garden absorbs.

Too often, though, we return all that carbon dioxide right back into the atmosphere by firing up a petrol-powered mower, filling pots with peat-based compost, or scattering artificial fertilizers. So, low-carbon gardening needs a twin approach. Lower your carbon emissions to neutral by gardening sustainably, with a light touch, then maximize the carbon your garden sequesters and stores, and help it actively combat climate change.

ABOUT THIS BOOK

The world of greenhouse gas emissions is mind-bogglingly complex for most scientists, never mind a simple gardener. Throughout this book, the focus is mainly on carbon dioxide emissions, but other

> *"A billion tiny actions have brought us to the edge of environmental crisis. And a billion tiny actions can pull us back from the brink."*

greenhouse gases such as nitrous oxide or methane have a part to play, too. So, I've occasionally adopted a widely used shorthand and use "carbon dioxide" as a catch-all to

------

THE UK'S MILLION ACRES OF GARDENS ALONE ALREADY LOCK AWAY **OVER 62 MILLION TONNES** (68 MILLION TONS) OF **CARBON**.

------

include several types of climate-changing greenhouse gases.

The frustratingly limited research into carbon sequestration and emissions in a domestic garden setting has also driven me to extrapolate many statistics from studies carried out in related but better-funded areas such as agriculture and forestry. So, such figures are rough equivalents rather than precise representations.

And, finally, not everyone will be able, or indeed want to take the "deep green" approach outlined on some of these pages. Gardening sustainably is a journey, and it's fine to start small and work up to the bigger changes later. The important thing is that everyone does something. A billion tiny actions have brought us to a point where the planet teeters on the edge of environmental crisis. And a billion tiny actions can pull us back from the brink.

HOW PLANTS
ABSORB CARBON

No matter how small your garden, no matter how humble, it has its tiny but important part to play in the great interconnected machinery that keeps our planet and its climate able to support life.

Plants rule the world. Pound for pound, they are by far the dominant life-form on Earth, accounting for about 80 per cent of the world's biomass (bacteria come a distant second at 15 per cent).

That's good news for us as we pump ever more greenhouse gases into the atmosphere, since plants can take them out again. Cars, aeroplanes, and factories burn fossil fuels for energy, releasing carbon dioxide into the atmosphere. But plants absorb, or sequester, carbon dioxide from the atmosphere in a process called photosynthesis, turning it into sugars. They then store it in woody stems, roots, and, later, decayed plant matter in soil, locking it out of the atmosphere for many years.

As gardens now offer the last safe refuge for plants in those areas of the world where humans have tidied away all other vestiges of nature, our plant-filled backyards have an increasingly significant role in combatting climate change.

THE CARBON CYCLE

Plants absorb carbon from the air, sequester it in their living tissues and then store it in the ground.

__The sun's energy is combined with carbon dioxide__ from the air by plants via photosynthesis. The plants emit oxygen and a tiny amount of carbon dioxide back into the air but absorb carbon and combine it with water from the soil to make the sugars, proteins, and carbohydrates they need to grow.

The earth's climate stays in equilibrium as long as the cycle is in balance, with as much carbon dioxide absorbed as released. Pump more carbon dioxide into the atmosphere than is absorbed, though, and the climate loses its delicately poised balance and starts to warm.

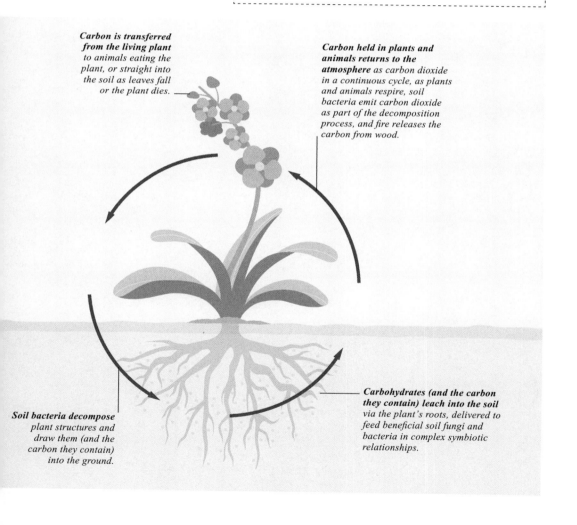

Carbon is transferred from the living plant to animals eating the plant, or straight into the soil as leaves fall or the plant dies.

Carbon held in plants and animals returns to the atmosphere as carbon dioxide in a continuous cycle, as plants and animals respire, soil bacteria emit carbon dioxide as part of the decomposition process, and fire releases the carbon from wood.

Soil bacteria decompose plant structures and draw them (and the carbon they contain) into the ground.

Carbohydrates (and the carbon they contain) leach into the soil via the plant's roots, delivered to feed beneficial soil fungi and bacteria in complex symbiotic relationships.

DESIGNING THE LOW-CARBON GARDEN

Low-carbon gardens absorb carbon, enhance biodiversity, and combat climate change. A little planning will help you maximize the positive impact of your garden.

THE LOW-CARBON GARDEN

A low-carbon garden buzzes with life, sparkles with water, and is packed with plants. If designed well, it can act as a carbon sink, actively combatting climate change.

1 Hedges *actively sequester and store carbon, as well as providing nesting sites, food sources, and shelter for wildlife.*
2 Fedges *will serve as somewhere to stack woody garden waste while it slowly rots down, providing shelter for wildlife, too.*
3 Coppices *of hazel trees provide a renewable, carbon-free source of beanpoles and pea sticks as well as a wildlife habitat.*
4 Vegetable gardens *store more carbon when permanently planted with fruit trees, berries, and perennial vegetables, interspersed with annuals to fill gaps.*
5 Compost bins *make good use of green waste from the kitchen and garden.*
6 Trees *suck in carbon, storing it in lignin-rich trunks and locking it in the ground.*
7 Wild meadows *of naturally occurring wildflowers are threaded with closely mown paths, which lead through the clouds of butterflies, bees, and other insects that thrive here.*
8 Tapestry lawns *are full of flowers and low-growing, mat-forming plants.*
9 Rain gardens and ponds *are designed to absorb excess rainwater; this densely planted damp garden also feeds into a deeper pond.*
10 Flower borders *line the central path, with layers of trees, shrubs, and perennials.*
11 Rainwater harvesting *will keep your dependence on mains water supplies to a minimum in summer.*
12 Driveways *of reclaimed stone slabs, infilled with low-growing lawn chamomile, provide parking but with added scent, beauty, and rainwater absorption.*

SMALLER SPACES

If space in your garden is restricted, any element from this design can be borrowed and adapted to fit a smaller garden.

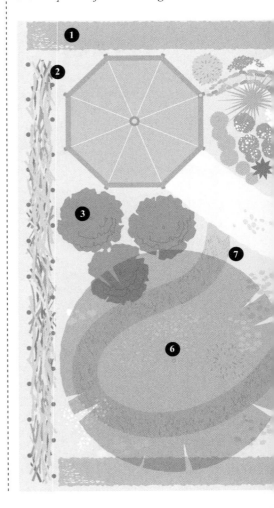

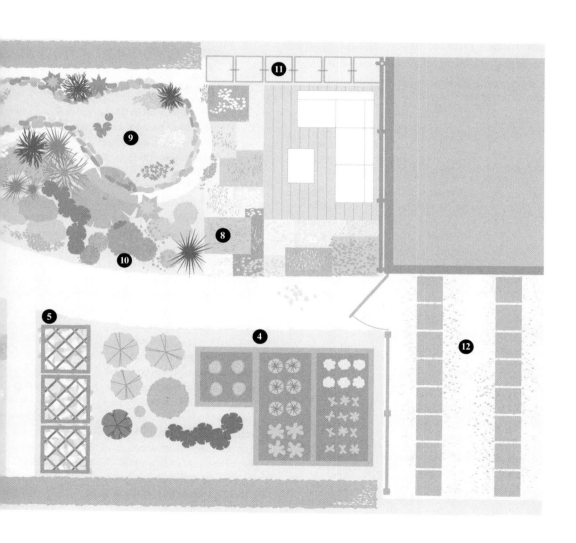

USING SUSTAINABLE MATERIALS

The materials you use in your garden can have a detrimental effect on the environment. Hard landscaping, especially when newly made and transported – sometimes from the other side of the world – can notch up some seriously high carbon emissions, undoing much of the good you do by planting.

The most commonly used materials are also the most carbon hungry. Cement – used to make concrete – contributes almost 1kg (2¼lb) of carbon dioxide for every 1kg (2¼lb) produced, although some of that is reabsorbed as concrete corrodes. Bricks add 250g (8oz) per 1kg (2¼lb), and every sq m (11sq ft) of stone patio adds about 47.5kg (105lb) of carbon dioxide, depending on the stone.

Wood is the only exception, as long as it's from managed forests. Look for the Forest Stewardship Council™ (FSC™) or Programme for the Endorsement of Forest Certification (PEFC) mark. New wood for building can be carbon neutral, so using it actively helps combat climate change.

CUTTING THE CARBON COST

A sustainably designed garden can still have patios and paths, but they're kept to a minimum and made from recycled ceramic "gravel" or reclaimed stone slabs. The sustainable garden has pergolas with untreated, locally sourced red cedar, and second-hand garden tables and benches, or handmade items in green oak.

Don't buy materials unless you have to, working with what you already have in the garden. Sourcing locally helps lower the high carbon cost of transporting heavy items. Second-hand or reclaimed materials from salvage yards carry a fraction of the carbon cost of new ones. They're also unique and lend your garden a timeless, classic look that's difficult to achieve with new materials.

AFTER WATER, **CONCRETE** IS THE MOST **CONSUMED MATERIAL** IN THE **WORLD**. EVERY YEAR, **3 TONNES** (3.3 TONS) IS USED FOR EVERY **PERSON** ON THE **PLANET**.

"New wood for building can be carbon neutral, so using it in the garden actively helps combat climate change."

KEEPING IT NATURAL

Treated softwood may be cheaper, but it's usually preserved with toxic chemicals or heat-treated (with all the carbon emissions that implies). Instead, use naturally durable timber: Western red cedar, green oak, and sweet chestnut all have a traditional look and will last for years. Or paint the wood yourself, with a mineral-based eco-friendly wood preservative.

CHOOSING PERMEABLE LANDSCAPING

How you use hard landscaping really matters, too. Large expanses of stone paving reflect heat, especially at night, adding to poor air quality and the "heat island" effect in cities. Stone paving also contributes to flash flooding by stopping water from filtering back into the ground. Keep paving to a minimum, use permeable materials like gravel, and maximize rainwater-absorbing planting to create a softer, more pleasing garden, which is gentler on the environment, too.

SUSTAINABLE ALTERNATIVES TO CONCRETE

The building industry is currently exploring greener alternatives to conventional cement-based concrete, made from waste materials or biodegradable fibres.

- **Ferrock** is undergoing field trials. It's made from 90 per cent waste materials, mostly iron dust, which reacts with carbon dioxide to create iron carbonate – therefore actively absorbing carbon from the atmosphere.
- **Timbercrete** is a reduced-cement concrete, substituting up to 10 per cent with sawmill waste to make a material that's about two and a half times lighter than concrete. It's available as blocks, bricks, and pavers.
- **Rammed earth** is compacted subsoil or chalk packed between temporary panels. You can use it for steps, walls, and even moulded furniture – but in damp climates it can be relatively short lived.

PLANTING THE LOW-CARBON GARDEN

A garden designed to absorb carbon is a plant-lover's heaven. The more plants you grow, the more carbon dioxide your garden can lock away – so it's the best excuse to pack in as many plants as you can.

Plants take in carbon dioxide during photosynthesis, much like we breathe oxygen. They then convert it into glucose, which helps them to grow. So when you put plants front and centre, keeping hard landscaping to a minimum, your garden becomes a living sponge, sucking carbon dioxide directly out of the air.

SLOW-GROWING **DECIDUOUS** TREES AND FAST-GROWING **CONIFERS** ABSORB ABOUT THE **SAME AMOUNTS OF CARBON**.

PLANT IN LAYERS

Think in canopies, like those of a forest. Layered planting puts taller trees overhead, underplanting them with shrubs, then perennials and ground cover, all densely planted to knit together and create a seamless tapestry of greenery.

PLANT PERMANENTLY

The more you plant and replant, the more you have to cultivate the soil,

disturbing fragile ecosystems and releasing the locked-in carbon underground. Focus on displays of long-lived, low-maintenance perennial plants that stay in place for many years.

LOVE YOUR SHRUBS!

Plants with a woody, permanent framework of branches lock up most carbon, but your choice isn't limited to trees. Shrubberies and hedgerows are really effective carbon sinks, and shrubs can often work well in gardens that are too small for trees.

WORK WITH YOUR GARDEN

Plants thrive when grown in the right conditions and will need fewer carbon-hungry inputs like fertilizer. Plant sun-lovers in the driest spots and site your pond where rain collects naturally in winter. And choose plants that won't outgrow the space, so you don't have to keep cutting them back.

DESIGN CLOSE TO HOME

Native and non-native plants absorb carbon equally well. But demanding, non-native plants that grow slowly and require high levels of fertilizer, watering, and extra heat may end up costing more in carbon emissions than they absorb – especially if they're imported too.

"Put plants front and centre and your garden becomes a sponge, sucking carbon dioxide directly out of the air."

GARDEN BUILDINGS AND STRUCTURES

When it comes to adding structures to your low-carbon garden, there are ways you can keep the carbon cost to a minimum, and at the core is careful material selection.

LOW-CARBON GARDEN BUILDINGS

Any structure you put in your garden is going to carry some sort of carbon footprint. But if you think a little flexibly when choosing materials and deciding on the design, you can keep the environmental cost to a minimum.

Green oak gazebo

Low-impact buildings are designed using natural materials, often made of plants, with high levels of "embodied" carbon (carbon held within the material instead of being released into the air). Production is low carbon too, because it avoids high-energy processes such as firing bricks at high temperatures. Eventually, at the end of its life, the building simply returns to the earth it came from.

GREEN OAK GAZEBO

It's possible to make a beautiful garden gazebo using traditional green oak carpentry, a British craft that dates back for centuries. The wood – used fresh cut – is still soft and easily sawn, and you can join it using wooden pegs and tenon joints instead of nails.

As it dries out after construction, the gazebo moves, splitting here and there, which adds to its character. A skilled green-oak carpenter will also use this movement to tighten the joints further.

Wood is one of the few carbon-negative materials you can use in the garden, as long as it's sourced locally. Over the lifetime of the tree that it's cut from, it absorbs and locks up more carbon from the atmosphere than the building process added. And a benefit of oak, which is naturally hard, is that it can be used outdoors untreated (as can chestnut and Western red cedar).

BUILDING A COB GARDEN HOUSE

Building with cob is one of the oldest building techniques – many cob houses in southwest England date back over 500 years. Cob builders mix earth, straw, sand, and water into a clay-like substance, which they then pack together to form immensely strong walls, up to 60cm (23½in) thick.

There's an old saying from Devon, England, which says that cob needs "a good hat and a good pair of boots", as keeping it dry is the main challenge. In return, you'll get a building that is naturally well insulated, warm in winter, and cool in summer. The walls absorb heat and release it overnight, creating a heated microclimate.

Cob garden house

HEMPCRETE SHED

Hempcrete is a type of concrete alternative made from plant fibres, combining fibrous stalks of hemp with lime. These alternative concretes store the embodied carbon contained in the plant material.

Hempcrete is available as blocks but can be moulded. It isn't load-bearing, though, so it will need a timber form to give it strength and shape.

Hempcrete shed

STRAW-BALE STUDIO

If you are considering adding a studio or office to your garden, building with straw bales is a durable and very cost-effective option – in fact, you can even build one yourself. Straw-bale building is a relatively recent phenomenon in the UK, though in the US there are people living in straw-bale houses that are more than 100 years old.

Straw is a waste material left over from barley or wheat harvests, and like all plant-based building materials, it locks up carbon within the walls. The bales are stacked like bricks, held in place with hazel pins, and timber boxes are built into the structure to hold doors and windows. Bales can also be packed tightly into a timber framework as infill. The whole structure is then rendered with lime or clay plaster. Studies have shown that straw-bale structures are remarkably resistant to fire.

FINISHING TOUCHES

Once you have chosen the main building material, you can add some of these low-impact finishes for roofs, renders, paints, and stains:

- **Roofs:** Thatch, turf, green, or cedar shingles.
- **Renders:** Lime, clay.
- **Paints and stains:** Water-based, limewash.

BUILD A RECLAIMED SHED

The carbon footprint of your garden building goes down even further if you can build it yourself from second-hand and reclaimed timber. Reusing wood keeps its embodied carbon locked safely away for many more years.

Wooden products are easy to recycle, but when you take an unwanted wooden door or pallet to the tip, it starts its journey on a high-carbon-emission process.

First, it's transported to a processing centre, then fed through a shredder before it's recycled into chipboard, animal bedding, or fuel for biomass boilers. All these are downcycled, lower-quality products – one step closer to final disposal when the carbon from the original tree is once again released back into the environment.

LOW-CARBON SHEDS

Most of this "waste" timber is perfectly reusable with hardly any extra reprocessing. Reclaimed wood has a fraction of the carbon footprint of new wood, as you don't spend all that energy chopping down a tree and turning it into planks. Plus, you get the satisfaction of producing a wonderfully quirky, characterful shed for your garden or allotment at a fraction of the cost.

FINDING RECLAIMED WOOD

- **Scaffolding yards:** Strict health and safety rules mean scaffolders have to discard boards showing signs of wear and tear – but they make wonderfully sturdy flooring for sheds.
- **Skips:** If you know a neighbour is planning a refurbishment and is pulling up wooden flooring or changing doors, ask if you can help yourself from their skip.
- **Pallets:** Local businesses often let you have their unwanted pallets for free, or you could look online. Look for locally sourced unstamped pallets as these are unlikely to be treated. On stamped pallets, IPPC means the wood has been treated to kill invasive insects and diseases; KD (kiln dried) and HT (heat treated) means the pallet is safe, but high carbon; MB means it's treated with toxic methyl bromide, so don't use it.
- **Community wood recycling enterprises:** Community projects reclaim timber then sell it on.

SHED BUILDING TIPS

- Check planning regulations with your local authority.
- Use reclaimed tiles, cedar shingles, or a planted "green" roof instead of bitumen-coated roofing felt.
- Use water-based varnishes and paints rather than oil-based ones.

MATERIALS FOR A RECLAIMED SHED

You first need to dig out the site and set reclaimed paving slabs level on a bed of sharp sand, one at each corner and one in the middle. Then make a grid from reclaimed timber to hold the floor and rest it on the paving slabs.

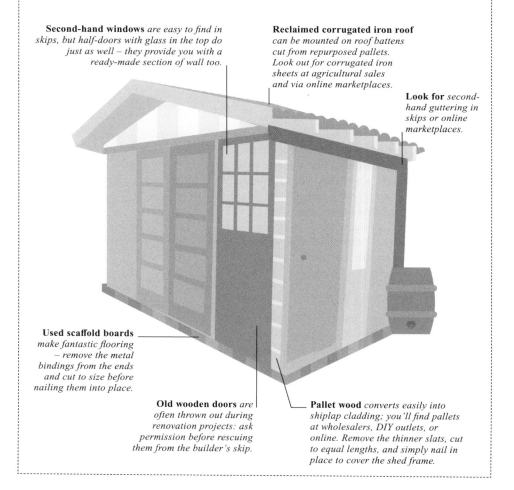

Second-hand windows *are easy to find in skips, but half-doors with glass in the top do just as well – they provide you with a ready-made section of wall too.*

Reclaimed corrugated iron roof *can be mounted on roof battens cut from repurposed pallets. Look out for corrugated iron sheets at agricultural sales and via online marketplaces.*

Look for *second-hand guttering in skips or online marketplaces.*

Used scaffold boards *make fantastic flooring – remove the metal bindings from the ends and cut to size before nailing them into place.*

Old wooden doors *are often thrown out during renovation projects: ask permission before rescuing them from the builder's skip.*

Pallet wood *converts easily into shiplap cladding; you'll find pallets at wholesalers, DIY outlets, or online. Remove the thinner slats, cut to equal lengths, and simply nail in place to cover the shed frame.*

GREEN YOUR SHED ROOF

Lower the footprint of your reclaimed shed even further by adding a green roof – plant it with herbs and fruit, or simply allow it to run wild with grasses and nectar-rich flowers.

Green roofs are heavy – plants plus damp compost can weigh over 100kg per sq. m (20lbs per sq. ft), so extra support is usually needed. For larger buildings, you will need to hire a structural engineer to check that the load-bearing capacity is up to the job.

If you're building your own shed, add extra roof beams and stronger walls. Keep the weight off bought sheds by sinking four sturdy pillars into the ground, one at each corner, and resting the green roof on top.

Build your green roof in layers, with a frame to hold everything in place. A waterproof membrane protects the roof, while a filter sheet prevents roots escaping.

WHAT TO PLANT

The simplest green roof is planted with sedums – drought-resistant succulents that can grow in just 8cm (3in) of substrate. Deepen that to 15cm (6in), though, and your roof can sequester more carbon – grasses and wildflowers, offering habitat and nectar to wildlife, or an edible green roof planted with thyme, prostrate rosemary, and alpine strawberries.

CARBON SINKS

A green roof transforms structures into mini-carbon sinks, taking your low-carbon garden to another level.

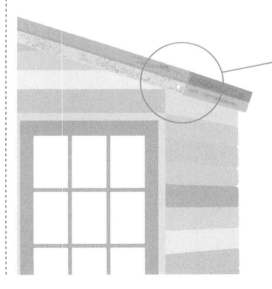

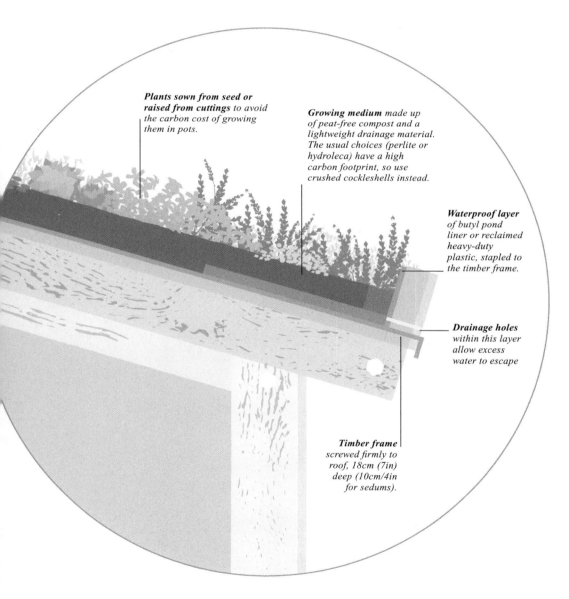

Plants sown from seed or raised from cuttings to avoid the carbon cost of growing them in pots.

Growing medium made up of peat-free compost and a lightweight drainage material. The usual choices (perlite or hydroleca) have a high carbon footprint, so use crushed cockleshells instead.

Waterproof layer of butyl pond liner or reclaimed heavy-duty plastic, stapled to the timber frame.

Drainage holes within this layer allow excess water to escape

Timber frame screwed firmly to roof, 18cm (7in) deep (10cm/4in for sedums).

PLANTING

Every plant actively extracts carbon dioxide from the atmosphere. Pack your garden with a variety of trees, shrubs, and perennials, and it can make an important contribution in the fight against climate change.

PLANT A TREE

Plant a tree in your garden and you help to offset not just your own carbon footprint, but that of your children and, with a bit of luck, their children too. It's one of the most effective ways you can put your garden to work in fighting against climate change.

Trees absorb carbon dioxide in massive quantities, converting it into rugged trunks and handsome branches that last for decades. The sheer amount of biomass (organic material) in a tree, both above and below ground, is one of the reasons they're so effective at sequestering (storing) carbon. It's estimated that a single broadleaf tree – an oak, say – stores 2.9 tonnes (3.2 tons) of carbon in its average 100-year lifetime.

Of course, not all of us have room in the garden for a spreading oak tree. But while it's true that the bigger the tree, the more carbon it stores, even small trees will live longer than most other plants you'll grow.

Small trees also add carbon continuously to the garden soil in the shape of leaf litter and deadwood. And they help your garden hold onto carbon simply by being permanently planted, so the soil is never disturbed and carbon stays locked safely underground.

CHOOSING TREES

Different trees absorb carbon at different rates. Tall, fast-growing trees such as poplars, willows, and silver birch will lock up carbon the most rapidly. But a slower-growing oak will eventually grow larger and live longer. Trees suited to your local climate (though not necessarily native species) will grow more vigorously and absorb more carbon.

WHERE TO BUY

Where to source your tree is an important decision. Trees bought from garden centres carry embedded carbon emissions from fertilizers,

IT'S ESTIMATED THAT **A SINGLE BROADLEAF TREE** STORES **2.9 TONNES (3.2 TONS)** OF **CARBON** IN ITS AVERAGE 100-YEAR LIFETIME.

> *"Planting a tree is one of the most effective ways you can put your garden to work in contributing to the fight against climate change."*

extra water, and transport. If those trees are also grown in peat-based compost and imported, their carbon footprint is even higher. In fact, the carbon cost of producing a tree commercially means a wait of three to 10 years after planting before that tree becomes carbon neutral.

Field-grown whips (very young trees) reduce the carbon footprint, especially if sourced locally. But the lowest carbon option of all is to raise your tree yourself. It's a satisfying gardening challenge, but it takes patience as it can be years before your tree comes to maturity. Many trees including horse chestnuts and oaks are easy to sow direct in autumn; and holly, London plane, willow, and poplar root well from hardwood cuttings taken in winter.

Choose from the following options, depending on the space you have.

LARGE GARDENS

- **Poplar** (*Populus nigra*)
- **English oak** (*Quercus robur*)
- **Tasmanian blue gum** (*Eucalyptus globulus*)
- **Yew** (*Taxus baccata*)
- **Field maple** (*Acer campestre*)

MEDIUM-SIZED GARDENS

- **Hazel** (*Corylus avellana*)
- **Goat willow** (*Salix caprea*)
- **Crab apple** (*Malus sylvestris*)
- **Callery pear 'Chanticleer'** (*Pyrus calleryana* 'Chanticleer')
- **Rowan** (*Sorbus aucuparia*)

SMALL GARDENS

- **Juneberry** (*Amelanchier lamarckii*)
- **Hawthorn** (*Crataegus laevigata*)
- **Apple** (*Malus domestica*)
- **Sour cherry** (*Prunus cerasus*)
- **Blackthorn** (*Prunus spinosa*)

CONTAINERS

- **Japanese maple** (*Acer palmatum*)
- **Cherry 'Kojo-no-mai'** (*Prunus incisa* 'Kojo-no-mai')
- **Bay tree** (*Laurus nobilis*)

CARBON OFFSET PLANTING

Trees are so efficient at extracting carbon dioxide from the atmosphere that planting them has become a recognized way of "offsetting" your carbon emissions in other areas of life. But it doesn't stop at trees – your whole garden can offset at least some of your carbon footprint.

It's impossible to arrive at an exact figure, but studies of urban green spaces suggest that mixed planting – planting a mixture of trees and plants close together – absorbs approximately 1 tonne (1.1 tons) of carbon per 0.4 hectare (1 acre) every year. Plant a wide mix of trees, shrubs, perennials, and ground cover, and your garden becomes a miniature carbon sink.

The types of plants you grow make a big difference to your garden's carbon offset potential.

PLANTS ABSORB UP TO **29 PER CENT** OF CARBON **EMISSIONS** PRODUCED BY **HUMAN** ACTIVITY.

- **Shrubs:** Woody, densely branched and long lived, shrubs are second only to trees in their ability to sequester carbon. Good choices are spindle (*Euonymus europaeus*), sweet briar (*Rosa rubiginosa*), and cinquefoil (*Potentilla fruticosa*).
- **Woody herbs:** Mediterranean herbs have one foot in the veg patch and the other in the shrubbery. They'll do wonders for your cooking too! Good choices are rosemary (*Salvia rosmarinus*), thyme (*Thymus vulgaris*), and sage (*Salvia officinalis*).
- **Fast-growing plants:** The quicker a plant grows, the hungrier it is for carbon to convert into the sugars that fuel all that burgeoning biomass. Good choices are clump-forming bamboo (*Chusquea, Fargesia,* or *Dendrocalamus*), figs (*Ficus carica*), and willow (*Salix alba*).
- **Grasses:** Most of the carbon in your garden is underground, and ornamental grasses have extensive, fibrous root systems as well as lots of above-ground biomass. Good choices are golden oats (*Stipa gigantea*), eulalia (*Miscanthus sinensis*), and pheasant's tail grass (*Anemanthele lessoniana*).
- **Long-lived woody perennials:** Digging holes releases carbon from the soil, so the longer your plant is in the ground, the better. Good choices are peonies (*Paeonia officinalis*), Balkan clary (*Salvia nemorosa*), and sunflower (*Helianthus* 'Lemon Queen').

"Plant a wide mix of trees, shrubs, perennials, and ground cover, and your garden becomes a miniature carbon sink."

COPPICING AND POLLARDING

In many ways a low-carbon garden is also a low-maintenance garden. You pack borders with shrubs and perennials and simply let them grow – after all, the larger your plant, the more carbon it absorbs, right?

Well, mostly, yes. But surprisingly, when you regularly cut back trees and shrubs to a woody stump, they react with a growth spurt and actually produce at least as much wood as if you hadn't pruned them.

A regularly coppiced tree (cut at ground level) can lock up roughly as much carbon as a fully grown one – and it will live longer, too, as coppicing keeps trees in a state of perpetual youth. The normal lifespan of a hazel tree is 80 years, but there is a coppiced hazel tree growing in Essex, in the UK, that is more than 500 years old.

SMALL SPACES

In the garden, pruning means you can still get the same carbon-offsetting benefits from a much smaller space. Eucalyptus trees, for example, are ideal for low-carbon gardens as they are fast growing and carbon hungry. But fully grown trees can top 30m (100ft), and most gardens would struggle to accommodate one. Coppice them, though, and they send up elegant, willowy stems with

beautifully rounded juvenile leaves, reaching 2–3m (6½–10ft) – infinitely more garden friendly, yet still absorbing just as much carbon.

WOOD AS A BIOFUEL

Most studies into carbon sequestration by coppiced trees and shrubs look at whether using them for biofuel (fuel developed from organic materials in place of fossil fuels) could be not only renewable but also a carbon-neutral energy source. The answer is yes, just about – but imagine what you could achieve if you didn't burn the stems you prune. Instead of releasing all the carbon they contain back into the atmosphere, you'd keep it within your garden.

GARDENING SUPPORTS

Use hazel rods for beanpoles and pea sticks and you save all the carbon you would spend going to the garden centre and buying imported bamboo. Weave coloured willow or dogwood into hazel rods to make zero-carbon climbing frames for sweet peas. And once you've finished using them, let them rot back into the soil, bringing the carbon they contain full circle.

HOW AND WHEN TO PRUNE

STOOLING

Cut back *to 10–15cm (4–6in) in early spring (mid-spring for dogwoods and willows). In subsequent years, prune to 5–8cm (2–3in) from the stubs.*

How often? *Every 1–2 years.*

Suitable for *smoke tree* (Cotinus coggygria), *shrubby dogwoods* (Cornus alba, C. sericea, *and* C. sanguinea), *willow* (Salix alba).

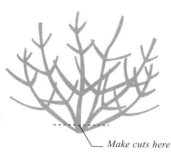

Make cuts here

Make cuts here

POLLARDING

Cut back *all branches to 5–8cm (2–3in) from the main stem in early spring, once the main stem reaches the desired height – usually 1.8–3.5m (6–11½ft).*

How often? *Every 2–4 years.*

Suitable for *lime* (Tilia x europaea), *London plane* (Platanus x hispanica), *foxglove tree* (Paulownia tomentosa), *Indian bean tree* (Catalpa bignonioides).

COPPICING

Cut back *hard to 15–20cm (6–8in) from the ground in late winter, while the trees are still dormant. In subsequent years, cut back to 5–8cm (2–3in) from the stump.*

How often? *Every 3–5 years.*

Suitable for *hazel* (Corylus avellana), *sweet chestnut* (Castanea sativa), *English oak* (Quercus robur), *cider gum* (Eucalyptus gunnii).

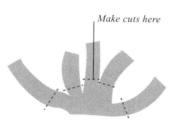

Make cuts here

MANAGING A LOW-CARBON MIXED BORDER

A light touch is all that's needed for a low-carbon mixed border. The idea is to lock as much carbon into the plants and soil as you can and keep it there for as long as possible.

Garden borders gently to keep carbon locked up – leave the soil undisturbed, and use spent stems and leaves wisely. As they rot, they transfer carbon deep into the soil. Once there, it is stored out of harm's way for years and boosts the soil so that your plants flourish.

SPRING

1 Clear away spent perennial stems and "comb" dead foliage out of evergreen grasses. Compost the waste or leave as a mulch on the soil.
2 Look out for self-sown seedlings to dig up carefully and replant to fill gaps between plants.
3 Top up mulches and add an extra layer of home-made compost to cover any bare patches of soil.

SUMMER

4 Put in plant supports over top-heavy plants such as peonies, weaving together slender branches of home-grown willow, dogwood, or hazel.
5 Chelsea-chop back summer- and autumn-flowering perennials, cutting back stems to encourage vigorous growth and more flowers.

MAINTAINING YOUR BORDERS

Follow these tips on "gentle gardening" to ensure your borders' success throughout the year.

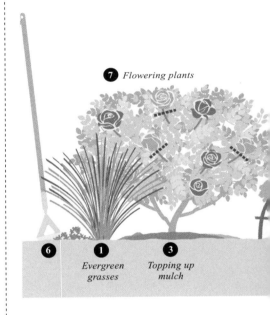

7 *Flowering plants*

6   **1** *Evergreen grasses*   **3** *Topping up mulch*

6 Hoe weeds instead of digging them out.
7 Deadhead to encourage more flowers, composting the spent flowers or letting them decay on the ground.

AUTUMN

8 Mulch for winter, adding a layer of home-made compost between plants and over any bare soil.

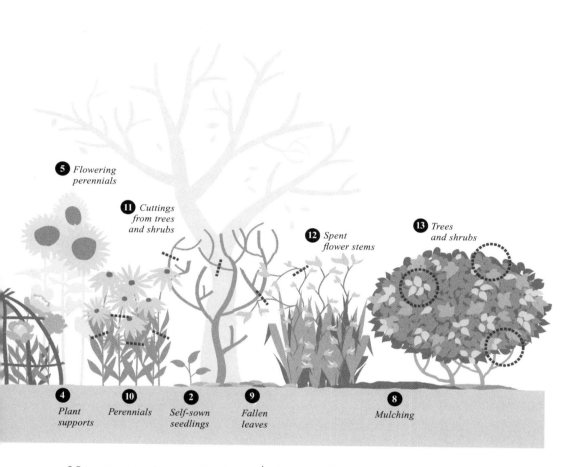

5 *Flowering perennials*

11 *Cuttings from trees and shrubs*

12 *Spent flower stems*

13 *Trees and shrubs*

4 *Plant supports*

10 *Perennials*

2 *Self-sown seedlings*

9 *Fallen leaves*

8 *Mulching*

9 Leave autumn leaves on borders, but clear any leaves that fall on plants where they may encourage disease.
10 Save seeds from favourite perennials such as aquilegia, primula, and rudbeckia for sowing in spring.

WINTER

11 Take hardwood cuttings from shrubs and trees and sink them into "slit" trenches made in the ground, burying them to three-quarters of their length.
12 Remove spent flower stems if they collapse, but leave as many on as possible for shelter for wildlife.
13 Carry out winter pollarding, coppicing, and stooling while trees and shrubs are dormant.

LAWN CARE

You don't have to give up your lawn to be a climate-friendly gardener – use sustainable lawn care practices and you will avoid the carbon emissions associated with intensive lawn management.

KEEPING YOUR LAWN GREEN

A lawn is much greener – literally and figuratively – than, say, concrete paving, but our lawns may be costing us dear in the battle against climate change.

When Edwin Beard Budding, a British engineer, invented his grass shearing machine back in 1830, he couldn't have guessed what a revolution he was unleashing on the world's gardens. The close-cropped lawn has been at the heart of our gardens ever since, and the gardening season is peppered with lawn-care tasks. We put a lot of love into our lawns – it's a great place for a picnic and it sets off our flower beds beautifully. Research has also shown that walking barefoot on grass helps to improve sleep, reduce pain, and lower stress.

EVEN NEW **PETROL** MOWERS SPEW OUT AS MUCH **POLLUTION** IN AN HOUR AS A **CAR** TRAVELLING **150KM (93 MILES)**, AS WELL AS EMITTING **GREENHOUSE** GASES.

You might think grass is good for the environment – most lawns do absorb some carbon dioxide and lock it into the ground. But studies have shown that the intensive management required for a well-kept lawn cancels out the carbon it absorbs.

THE CARBON COST

If you mow your lawn weekly with a petrol mower, fertilize it, and water it regularly, your lawn is responsible for considerably more carbon emissions than it's capable of absorbing. You can reduce those emissions by up to 70 per cent – and bring your lawn much closer to carbon neutral – just by relaxing a little on many fronts.

Instead of artificially fertilizing your lawn, feed it by leaving the nitrogen-rich grass clippings on the surface. Tolerate and learn to love low-growing weeds such as daisies and clover – they will absorb more carbon than grass, provide nectar for bees, and keep your lawn looking greener in droughts, so you can put away that hosepipe.

Among the most costly parts of your lawn-care routine in terms of carbon emissions, though, is the mowing. Even brand-new petrol mowers spew

> *"The development of lithium-ion rechargeable batteries has ushered in a new generation of battery-powered lawnmowers."*

out as much pollution in an hour as a car travelling 150km (93 miles), along with potent greenhouse gases such as nitrogen oxide. And that's before you factor in the carbon emissions of extracting fossil fuels or manufacturing the mower.

GREENER OPTIONS

Until recently, apart from push mowers (the greenest option, but only really practical for very small lawns), the only alternative to petrol has been electric mowers. The energy they use can be low carbon if you source your electricity from renewable energy supplies, but they have the big disadvantages of cables that limit your reach and wrap around your ankles at inconvenient moments.

Now, though, the development of lithium-ion rechargeable batteries for electric cars has also ushered in a new generation of battery-powered lawnmowers. These are cordless and powerful enough to tackle large and overgrown lawns – yet without emitting pollution. Mow your lawn

with one of these, perhaps a little less often, and your lawn is on the way to being greener.

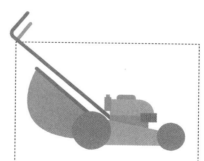

LITHIUM-ION BATTERIES

Even if you mow with a rechargeable battery-powered mower, your lawn still won't be totally green. Manufacturing the batteries involves high-carbon emissions to fracture rock during lithium mining. And less than 5 per cent of lithium-ion batteries are currently recycled, making them a non-renewable technology. They still beat fossil fuels hands down, though.

LAWNS WITHOUT THE GRASS

Although grass absorbs some carbon, low-growing, woody-stemmed perennials absorb much more. No-grass lawns have the benefits of a grass lawn but without the eco-drawbacks of weekly mowing.

No-grass lawns use mainly low-growing perennial plants, which reproduce clonally (where only one plant is involved). The plants spread over the ground through underground stems, adventitious roots (originating from stems), or above-ground runners to form dense mats of greenery. The effect is a sheet of plants you can walk or sit on, just like a conventional lawn but with added scent, flowers, and carbon sequestration.

Lawns made from low-growing plants instead of grass, suit smaller areas that aren't heavily used. They won't cope with a daily game of football with the kids, but they're a lovely choice for a small back-garden lawn for sitting out on and walking over once or twice a day.

PREPARING YOUR SPACE

Prepare the ground well for your no-grass lawn, removing weeds and large stones, and raking it level. You will need about 100 plants per sq m (11sq ft), spaced 10cm (4in) apart. You can sow flowering plants from seed, but it's best to buy named varieties as rooted runners (offsets that have been planted, rooted, and dug up for resale). You could also buy larger plants and split them into smaller sections.

Once planted, leave your lawn to grow for at least 12 weeks before walking on it, and keep weeding it while it grows. Grass-free lawns don't need fertilizing or regular mowing – simply shear off top growth once or twice a year.

GRASS-FREE LAWN OPTIONS

- **Chamomile** (*Chamaemelum nobile*): The cultivar 'Treneague' is best for chamomile lawns as it doesn't flower, so it stays compact. Chamomile needs a sunny site and free-draining soil.
- **Creeping thyme** (*Thymus serpyllum*): Thyme lawns look spectacular when they are in flower, and thyme enjoys similar conditions to chamomile.
- **White clover** (*Trifolium repens*): You can sow clover from seed, though it needs reseeding after two to three years. New strains, known as "microclovers", may last longer.
- **Corsican mint** (*Mentha requienii*): This fragrant herb thrives in shade, but like all mints, it is invasive, so add a solid edge to your lawn to prevent it from escaping.

CREATE A TAPESTRY LAWN

A tapestry lawn is a no-grass lawn with many plant species grown together. The idea was pioneered by Dr Lionel Smith, a professor of ecology at Reading University in the UK, for a project sponsored by the Royal Horticultural Society (RHS).

1 *Choose at least 12 different plant varieties with a range of contrasting leaf colours and flowering times.*

2 *Plant in blocks with a 10cm (4in) space between the plants.*

3 *Water your lawn for the first month and remove any weeds that appear.*

4 *Start mowing once plants reach 6–9cm (2½–3½in) tall. Set blades to 4–5cm (1½–2in) and drive straight over the top – this doesn't harm the plants and keeps an even balance between taller and lower-growing varieties. Repeat the mowing three to five times each year.*

TAPESTRY LAWN PLANTS

- **Purple spineless acaena** (*Acaena inermis* 'Purpurea')
- **Yarrow** (*Achillea millefolium*)
- **Bugleweed** (*Ajuga reptans*)
- **Daisy** (*Bellis perennis*)
- **Wall bellflower** (*Campanula portenschlagiana*)
- **Chamomile** (*Chamaemelum nobile*)
- **Maiden pink** (*Dianthus deltoides*)
- **Hairless leptinella** (*Leptinella dioica*)
- **Snowdrop wood-sorrel** (*Oxalis magellanica*)
- **Mouse-ear hawkweed** (*Pilosella officinarum*)
- **White clover** (*Trifolium repens*)
- **Germander speedwell** (*Veronica chamaedrys*)

A WALK ON
THE WILD SIDE

You don't have to replace your lawn to turn it into a powerful carbon sink that can play a significant part in combatting climate change. All you have to do is stop mowing. Waiting beneath your clipped lawn is a riot of wildflowers and prairie-like grasses that will burst into life as soon as you let them.

Within a couple of weeks of letting your lawn grow naturally, clover and daisies will begin to spangle the smooth, monochrome green. Give it a little longer, and you'll start to spot more wildflowers, as well as bees and butterflies flocking to enjoy their nectar.

NATURAL GRASSLAND CAN LOCK UP OVER **3 TONNES (3.3 TONS)** OF CARBON PER **HECTARE** (2.5 ACRES).

SUPPORTING LIFE

When you put away the mower altogether, your lawn reverts to natural grassland – one of the world's most efficient carbon sinks, able to lock up over 3 tonnes (3.3 tons) of carbon per hectare (2.5 acres). As well as a huge diversity of wildflowers, you are providing food and habitat for insects to bring your garden to life.

WILD LAWN CARE

Studies have shown that the best approach to rewilding a lawn is a "mohican" cut, where you mow areas of your lawn at different rates, letting patches of grass grow longer.

- **Paths:** Mow paths weekly to keep them neatly cut through your wild lawn, ideally with a non-powered push mower. Add a clearing, too, so you can get up close to any wildflowers that grow.

- **Close to the house:** In places where you want your grass to stay tidy, mow once every four weeks. This produces the largest number of flowers and 10 times the amount of nectar for bees. Dandelions, white clover, selfheal, and buttercups will all grow.

- **Further from the house:** Leave these areas unmown all year and taller, slower-growing wildflowers will appear. Mowing annually produces a diversity of species such as ox-eye daisies, field scabious, knapweed, and campion. Longer grasses also take extra carbon deep into the soil.

"Put away the mower and your lawn reverts to natural grassland – one of the most efficient carbon sinks."

THE LOW-CARBON MEADOW

One of the loveliest of semi-natural grasslands is the flower-filled hay meadow – agricultural grassland left unmown to produce hay. Inside our gardens, too, it's possible to create a patch of meadow that could be the last refuge for some of our rarest and most beautiful species.

Traditionally managed hay meadows are left to grow through spring, then the hay is cut in summer. This allows wildflowers to flourish into a carpet of meadowsweet, lady's bedstraw, ragged robins, and cranesbills.

Hay meadows are now an increasingly rare sight. In the UK, for example, about 97 per cent of them have disappeared since the 1930s, when farmers ploughed to make way for growing crops, or "improved" them with fertilizers until the grass grew thick and ousted the wildflowers.

SMALL GARDEN MEADOWS

If you allow them space to grow, you'll find swathes of wildflowers will flourish. And as our gardens fill with wildflowers, they come alive. Meadows are among the best of wildlife habitats, with nectar-rich flowers and undisturbed, tussocky grass to shelter solitary bees and ground beetles. They attract bees, butterflies, and other pollinating insects, with many, like marbled white and meadow blue butterflies, found nowhere else.

THE LOW-CARBON WAY

Making a meadow the conventional way can be an intensive process. You'll need to use machines to strip off the nutrient-rich top layer of the soil, releasing much of the carbon within it back into the atmosphere. Buying seeds or plug plants (seedlings ready to plant), which are commercially produced off-site, carries a carbon cost too.

A low-carbon meadow, on the other hand, adds almost nothing to your garden's carbon footprint, as it makes the most of the wildflowers already growing in your garden. You probably know them as weeds, but buttercups, clovers, and dandelions are highly valuable food plants for bees and other pollinating insects.

WILDFLOWER BEAUTY

Once the grasses retreat, you'll find the first wave of more common wildflowers is joined by other, spectacular wildlings, such as knapweeds, campions, or lady's smock – and maybe, if you're exceptionally lucky, an orchid. Discovering new blow-ins becomes a constant delight. It's the meadow that nature would have created, tailor made for your garden.

"A low-carbon meadow adds almost nothing to your garden's carbon footprint."

HOW TO CREATE A LOW-CARBON MEADOW

1 *Scalp your grass in early autumn with the mower blades on their lowest setting. This weakens the grass and exposes occasional bare patches of soil. Remove all clippings.*

2 *Sow hemi-parasitic wildflowers onto the bare soil; once established, they weaken grasses, helping wildflowers dominate. Suitable choices include yellow rattle (*Rhinanthus minor*) and eyebright (*Euphrasia nemorosa*).*

3 *Don't mow throughout spring and summer. Large natives such as cow parsley, thistles, and nettles are fantastic for wildlife, providing food and shelter, so tolerate them as far as possible but remove selectively, by hand, if they get invasive.*

4 *Take an annual cut in late summer and rake up the "hay" to remove and add to compost. It may take some years to achieve a pleasing balance of grasses and wildflowers, but the result will be a long-lasting, well-established meadow.*

HEDGES AND BOUNDARIES

Hedges not only provide boundaries and shelter to your garden but also are rich in woody biomass. A well-grown hedge has the potential to sequester and store significant amounts of carbon.

PLANT A HEDGE

Hedges are the softest backdrop to a garden, like hanging a dark green curtain behind your plants so they can really shine. They're full of life, acting as green, leafy "apartment blocks" for birds, mammals, amphibians, and insects. And they soak up carbon like no other boundary.

A well-grown hedge is incredibly biomass-rich. It's full of carbon-holding wood and is densely branched, multiplying its capacity to absorb carbon dioxide still further.

Hedges also shelter your garden, filtering the wind and helping prevent soil erosion, as well as providing a welcome sound barrier. Hedges also absorb harmful pollution from vehicle emissions, and thorny ones make impenetrable security boundaries, too.

A 100M- (330FT)-LONG **MATURE, MIXED HEDGE** CAN ABSORB ABOUT **1.2 TONNES (1.3 TONS)** OF **CARBON DIOXIDE** A YEAR.

WHICH HEDGE TO PLANT?

Evergreen hedges are most efficient at absorbing carbon dioxide as they photosynthesize year round. Wider hedges are more efficient at sequestering carbon than taller ones and need less clipping. Slower-growing hedging plants need clipping less often – good for you and for the planet, as the energy used for trimming raises your carbon footprint.

- **Yew** (*Taxus baccata*): Dark green, glossy, and beautifully clippable, yew is the go-to hedge for formal shapes, and birds love the berries.
- **Darwin's barberry** (*Berberis darwinii*): Small dark green leaves, orange flowers, and purple berries make barberry an attractive hedge.
- **Japanese holly** (*Ilex crenata*): Tiny leaves and a dense growth habit make this a neat choice for low-growing hedges up to 1m (3ft).
- **Bay laurel** (*Laurus nobilis*): Often grown as a kitchen herb, bay also makes a lovely hedge with fragrant leaves and a dense habit.
- **Pittosporum** (*Pittosporum tenuifolium*): Handsome and elegant, black-stemmed and variegated varieties.

MANAGING YOUR HEDGE

- **Clip once a year** in late summer, after birds finish nesting; cut the top narrower so light reaches the base.
- **Cut by hand** using zero-emission, well-sharpened hedging shears.
- **Use electric or battery-powered** hedgetrimmers, powered with electricity from a renewable source.

"Hedges absorb harmful pollution from vehicle emissions."

PLANT A HEDGE FOR WILDLIFE

A single-species hedge goes a long way towards absorbing carbon. But you can double down on this by planting a mixed native hedge that grows into one big haven of blossom, foliage, and fruit.

Mixed hedges are not for neat freaks – you won't get perfect corners or razor-sharp edges, as the various plants grow at different rates. But they will soften your garden's boundaries and add endless seasonal interest, from clouds of hawthorn blossom in summer to rich, purple sloes in autumn. Include cherry, plums, blackberries, and wild pears, and you'll have fruit to pick, too.

WILDLIFE VISITORS

Mixed hedges are one of the best habitats you can create for wildlife – they provide safe "corridors" for voles and hedgehogs to move along, and berries for birds. Some butterflies could not survive without them. Brown hairstreak caterpillars feed only on blackthorn, for instance, which is found mainly in hedgerows.

NATURAL CARBON SINKS

Mixed hedges are likely to be better carbon sinks than formal hedges. In fact, one study found that forests store 6 per cent more carbon with each new type of tree introduced –
reinforcing the idea that mixing species is better for carbon sequestration. And mixed hedges certainly score well on sheer biomass – they spread up to 1.5m (5ft) wide – and the wider the hedge, the more carbon it locks up.

PLANTING A HEDGEROW

Even in a mixed hedgerow, there's usually one dominant species of tree. Aim for about 60–70 per cent of your hedge to be one type. Hedges are often hawthorn, but hazel, beech, or holly trees work well too.

Then add four or five other plant species, planting randomly in groups of two or three so they aren't overwhelmed by more vigorous plants. Once your hedge is established, add in dog roses, honeysuckle, or blackberries to thicken the foliage and add to its biodiversity.

PLANTS FOR A NATIVE MIXED HEDGE

- **Blackthorn** (*Prunus spinosa*)
- **Cherry plum** (*Prunus cerasifera*)
- **Elder** (*Sambucus nigra*)
- **Field maple** (*Acer campestre*)
- **Guelder rose** (*Viburnum opulus*)
- **Hawthorn** (*Crataegus monogyna*)
- **Holly** (*Ilex aquifolium*)
- **Hazel** (*Corylus avellana*)
- **Pear** (*Pyrus communis*)
- **Spindle** (*Euonymus europaeus*)

HOW TO PLANT A
WILDLIFE HEDGE

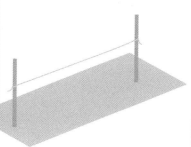

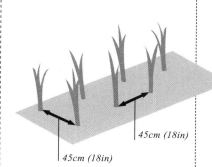

45cm (18in)

45cm (18in)

1 *Clear a strip of ground about 1m (3ft) wide, removing weeds and large stones, and set up a string line between two canes to give you a straight line to guide the planting.*

3 *Plant in staggered double rows, spacing plants 45cm (18in) apart. Use a spade to make a slit for each plant, then find the point where your plant's roots flare out from the stem and sink the plant in until this point is at soil level.*

2 *Buy hedging plants in winter when they're around two years old and 60cm (2ft) high. These should be bare root – supplied while not actively growing. Or, if you can wait, most native hedging plants are also easy to grow yourself from seeds or cuttings.*

4 *Fill in the soil, water, and then mulch with garden compost. Keep watered through the first year and protect from rabbits with a 1.2m (4ft) wire fence. Then sit back and enjoy the haven you've created for local wildlife.*

MAKE A FEDGE

Garden hedges – even wildlife hedges (see pp.50–51) – need some maintenance, even if only a rough cut in late summer to tidy them up. And that comes with its own carbon cost in the shape of powered hedge trimmers, plus disposal of woody trimmings. A dead hedge, or "fedge", on the other hand, looks after itself.

A fedge is somewhere between a hedge and a fence – it's a shaggy, natural-looking boundary, just like a hedge, but it's really a fence built from waste wood. It acts as a tall compost heap – but for all your woody prunings, the kind of garden waste you'd normally take to the tip.

NO MORE WOOD WASTE

Everything can go in your fedge – hedge trimmings, rose prunings, old branches, spent raspberry stems, dead brambles – anything too woody to go on a compost heap. Be sure to leave out material infected with serious diseases like canker or fire blight.

All this woody waste is crammed with carbon, which would only go back into the atmosphere if you chipped or, worse, burned it. Stack it in your fedge and it's held right where it is, sometimes for years, until it rots gently back into the soil. It's the ultimate in low-carbon gardening.

CREATE A FEDGE

Fedges are an excellent choice for informal garden boundaries. Shorter than living hedges, these beautiful woven borders can provide windbreaks, and a habitat for wildlife.

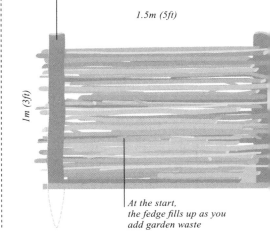

Drive stout, round poles *1.8m (6ft) long, firmly into the ground in double rows – about 1m (3ft) apart and 1m (3ft) high. Space the poles about 1.5m (5ft) apart to make two staggered lines. Ideally, cut your poles from five-year coppiced hazel grown in the garden.*

1.5m (5ft)

1m (3ft)

At the start, the fedge fills up as you add garden waste

Top view of a fedge, showing the staggered poles

Keep adding It'll take a few years for the first layer to break down fully (depending on the thickness of prunings), joining decayed leaf litter and other debris to make a layer of rich, brown organic matter. Top up with more waste as levels start to drop.

Tread down your fedge each time you are ready to add more. Compact it evenly to help the rotting process and continue to build up the sides with longer, thicker branches, adding shorter, twiggier prunings to the middle.

Add more on top as the prunings break down and the level of the dead hedge drops. This layer is the richest in wildlife, especially toads, slow worms, hedgehogs, and stag beetles.

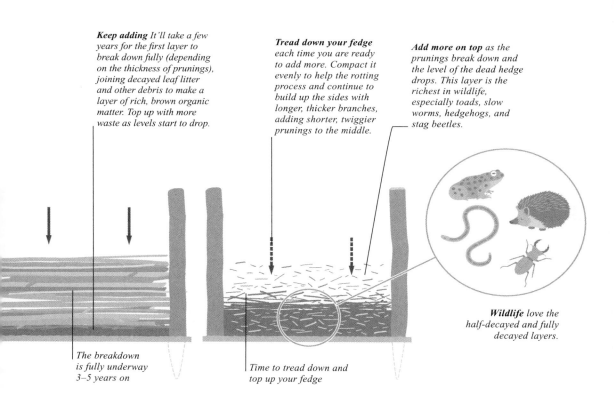

The breakdown is fully underway 3–5 years on

Time to tread down and top up your fedge

Wildlife love the half-decayed and fully decayed layers.

PATHS, PAVING, AND PATIOS

While paths, paving, and patios help us to enjoy our outdoor spaces, these hard surfaces can prevent natural drainage and come with a significant carbon cost. Discover other, simple solutions.

PLANTING
BETWEEN PAVERS

Most of the time, we inherit the paving we're given. When we move into a house, it's likely there's a patio already installed, and probably some paths here and there too. Unless there's a good reason to replace them, the best thing for your carbon footprint is to make the most of what you've got.

There's plenty you can do to improve the carbon impact of existing hard landscaping. Planting beside and into expanses of pavement absorbs excess rainwater for a fraction of the carbon emissions you incur in manufacturing, transporting, and installing permeable paving. Growing plants take in carbon and lock it in the soil. And besides, given the choice between thyme flowers and cement grouting, there is not much competition in terms of beauty.

PLANTING AROUND PAVING NOT ONLY PROVIDES **FOOD** AND **SHELTER** FOR **WILDLIFE**, IT ALSO ATTRACTS **POLLINATORS**, INCREASING THE **BIODIVERSITY** IN YOUR **GARDEN**.

PLANT THE JOINTS

If your existing path or patio is looking the worse for wear, with cracked and worn joints filling up with weeds,

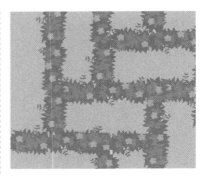
Plant your joints

don't repoint it with cement-based mortar– just plant into the gaps instead. The strips of plants mesh together to outcompete the weeds, and bursts of flowers will perfume the air as you walk on them. Plant with creeping thyme (*Thymus serpyllum*), Corsican mint (*Mentha requienii*), or Heath pearlwort (*Sagina subulata*).

STEPPING STONES

Take it a little further and you can widen the gaps to create stepping stones. Place individual slabs where

Stepping stones

your feet fall naturally. Lay the stones on a 1.5cm (⅝in) bed of sand so the tops are at ground level and plant in between. Plant with: chamomile (*Chamaemelum nobile*), bugle (*Ajuga reptans*), or New Zealand bur (*Acaena microphylla*).

A RIBBON DRIVEWAY

Planting a strip down the middle of your driveway can soften the impact of this area, blending it seamlessly into the wider garden. Planting directly alongside paved areas also creates a natural soakaway – a method of drawing surplus rainwater back into the earth. This is especially important for driveways, which usually feed directly out onto a public road. Plant with Mexican fleabane (*Erigeron karvinskianus*), trailing bellflower (*Campanula poscharskyana*), and grey cranesbill (*Geranium subcaulescens*).

Ribbon driveway

CREATE PATIO PLANTING POCKETS

Lift paving slabs to create ready-made planting areas, for anything from grasses and perennials to herbs. Once you remove the slab, dig out the sub-base, then refill with compost and topsoil. Plant with: strawberries, rosemary, and *Hebe* 'Red Edge'.

Patio planting pockets

HOW TO PLANT THE GAPS

1 *Scrape out old joints with a hooked patio weeding knife. Chisel out any stubborn bits with a bolster chisel and hammer.*
2 *Mix three parts multipurpose compost to one part sharp sand. Pour the mix over your paving and brush it neatly into the cracks.*
3 *For named varieties, split a large plant into several small sections, each with roots attached. Plant into the gaps about 20cm (8in) apart, backfilling with compost and firming in as you go.*
4 *You can sow creeping thyme from seed. Sprinkle the seed directly onto damp compost then press it in lightly.*

PERMEABLE PAVING

Paths, paving, and patios help us to enjoy the plants we've worked so hard to grow. To keep your carbon footprint low, make sure every piece of paving is there for a good reason and use permeable paving when you can.

Hard landscaping can provide you with somewhere to park the car and doesn't need much maintenance (unlike, say, a lawn). Paths, paving, and patios also lead us through our flowerbeds and give us a place to sit, soaking up the beauty. Overdo it, though, and hard landscaping can be the most expensive item in your garden – for your purse and your carbon footprint.

REDUCING RUN-OFF

Stone, concrete, and tarmac absorb up to 50 per cent less rainwater than most types of soil. In a severe storm, that's a lot of water running straight off your garden into the street drains – making it more likely that they will become overwhelmed, back up, and flood your house and garden.

Globally, the number of floods has surged since 1980 because the soil that should be absorbing all that water is disappearing under ever-expanding cities. When floodwaters wash away topsoil, the carbon it stores is released, and drowning plants also release methane – a greenhouse gas that is 30 times more efficient at trapping heat than carbon dioxide.

So, do source your materials carefully, using local and reclaimed stone wherever you can. And when you do need to include a larger area for a patio or car parking, use permeable paving. Letting the ground breathe allows rainwater to drain into the ground to be absorbed into the soil, just as it would if you were planting in that area.

TYPES OF PERMEABLE PAVING

- **Permeable block paving:** These are conventional blocks that are laid on grit to leave porous gaps through which surface water can pass. The water then drains through the hardcore (aggregate) back into the soil. Blocks made from mostly recycled concrete cut the carbon cost, as do reclaimed brick pavers.
- **Cellular paving:** These plastic or concrete grids are strong enough to take the weight of a car. Lay them on a sub-base of aggregates (such as sand and stone scalpings) and fill them with gravel or grass. Remember that though using grass as a filling will sequester more carbon, some of that is offset by the regular mowing required to keep it looking neat.

REDUCING RUN-OFF

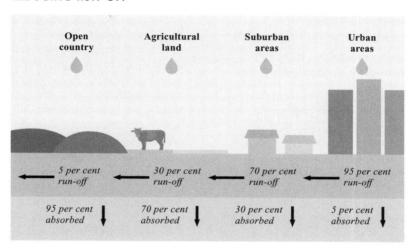

Open country	Agricultural land	Suburban areas	Urban areas

5 per cent run-off

30 per cent run-off

70 per cent run-off

95 per cent run-off

95 per cent absorbed

70 per cent absorbed

30 per cent absorbed

5 per cent absorbed

Plants help the ground absorb rainwater, reducing the likelihood of flooding.

- **Filling joints with sand:** Reclaimed stone paving slabs are a low-carbon way to make a patio. They're usually sealed with impermeable, cement-based mortar, so use dry sand to fill around the slabs instead and excess water will run straight through. Use specialist jointing sand and make sure it, and the paving, are completely dry before brushing it into the joints.
- **Gravel:** Source gravel carefully – if it's been dredged from the seabed, it can do untold environmental damage to marine life. Easy to install, reclaimed ceramic "gravel" repurposes old sinks and baths, crushing them into aggregates for driveways. Smaller pea gravels can scatter and are difficult to negotiate with a pushchair or wheelchair.
- **Self-binding gravel:** More expensive, but better behaved than conventional gravel, this option includes sand-like particles that, when wet, form a stable base with just a little loose material on the surface. It's usually locally quarried, but must be kept dry until laid, so be sure to use it straight away.

PATHS WITHOUT
THE PAVING

Garden paving is not compulsory. In fact, the less hard landscaping you have, the closer to zero carbon your garden is likely to be. Low-impact paths help you tread more lightly through your garden.

Low-impact paths don't require stone quarrying, cement, or long-distance lorry journeys. Everything is renewable, recyclable, and biodegradable – and pleasingly cheap. Organic paths do need more maintenance, though, and this can push their ultra-low carbon price back up if you're not careful.

THE **CEMENT** INDUSTRY PRODUCES **8 PER CENT** OF THE WORLD'S **CARBON DIOXIDE** EMISSIONS.

ORGANIC OPTIONS

Source materials for your low-impact paths locally, and keep maintenance low-tech by using a push mower (for grass paths) and weeding by hand.

- **Grass paths:** These paths are the easiest to create – simply define the edges with a traditional half-moon lawn edger. The carbon cost can be high, though, as you need to mow them every couple of weeks, and they harbour slugs and can be muddy.

- **Trodden-earth paths:** Strip away surface vegetation, then compact the soil with an earth rammer (a hand tool for compressing earth). Weeds are rarely a problem when heavily used, but this type of path can get muddy.
- **Mulch paths:** Lay thick cardboard, then cover with a mulch and tread down. Anything organic works: grass clippings, hay, sawdust, even garden prunings cut small. Top up constantly to keep weeds at bay.
- **Woodchip paths:** Contact your local tree surgeons to see if they can drop off a load of chipped wood. Woodchip makes an excellent path material, laid over cardboard. Top up once a year.
- **Log slices:** Repurpose felled trees into log-slice stepping stones or mosaic pathways. Check the logs haven't been treated with artificial chemicals first. Cut logs into 5–8cm (2–3in) thick rounds, paint with water-based wood preservative, then lay on top of aggregate. Fill gaps with low-growing herbs.
- **Boardwalks:** Recycle pallets into boardwalks. Raise them on horizontal bars or simply lay them flat on the ground. Cut boards to 75cm (30in) long, drill holes in each end, and peg in place with long nails. Replace any rotting boards each year.

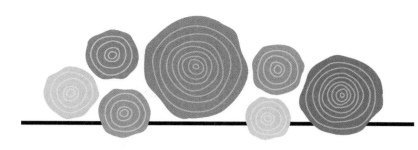

"Low-impact paths don't require quarrying or cement. Everything is renewable, recyclable, and biodegradable."

GREENHOUSE GROWING

Gardeners love greenhouses, but what is the true cost of protecting our plants? With some resourcefulness and ingenuity, climate-friendly gardeners can find ways to have a greenhouse while reducing its carbon cost.

THE GREENHOUSE EFFECT

Every gardener wants a greenhouse. It gives the few degrees of extra warmth you need to grow frost-tender tomatoes, overwinter pelargoniums, or sow seeds in early spring. And it's your little refuge, somewhere you can escape to and garden even in a downpour.

Unfortunately, greenhouses are carbon greedy. Everything growing in them needs extra food and water and, if heated, energy too.

The highest carbon emissions are in the building itself. Processing 1kg (2¼lb) of new aluminium emits an average 11.5kg (25lb) in carbon emissions. However, if you use recycled aluminium, this falls to just a little over 1kg (2¼lb) of carbon per 1kg (2¼lb) of frame. Wooden frames are more expensive, but timber actively stores carbon – so all you have to worry about is the glass, at up to 8.4kg (18½lb) of carbon dioxide emissions per kg (2¼lb), falling to 1.2kg (2½lb) if recycled after use.

SECOND-HAND SAVINGS

You can sidestep all that damage to the climate and still have a greenhouse, though – just acquire somebody else's. Unwanted greenhouses pop up in online auctions, freecycling sites, or the small ads.

COOLING EFFECT

Once you've installed your second-hand greenhouse, use it to lower carbon emissions by growing food, especially out of season. Restocking greenhouse borders with salads through winter requires no heating, no fertilizers, and little extra water. That's a big improvement on buying your salads from supermarkets, carrying high emissions from chemical pesticides and fertilizers, transport, cold storage, and single-use plastic packaging.

PROCESSING 1KG (2¼LB) OF NEW **ALUMINIUM** TO MAKE A GREENHOUSE **FRAME** EMITS AN AVERAGE **11.5KG** (25LB) IN **CARBON EMISSIONS**.

"Keeping your greenhouse even minimally frost free uses so much energy you risk frittering away your efforts to save carbon elsewhere in the garden."

LOW-CARBON HEATING

If you do want to heat your greenhouse, you're probably sheltering plants that can't cope with frost, such as citrus, pelargoniums, and rocoto chillies. But keeping your greenhouse even minimally frost free uses so much energy you risk frittering away your efforts to save carbon elsewhere in the garden. So think hard about whether you need to, and if so, do it as sustainably as you can. There are a few options.

- **Do without heating:** Wet kills more plants than cold, so water frost-tender plants sparingly and they'll often cope with a couple of degrees below zero. Below this, insulate individual plants temporarily overnight using newspaper or hessian (also known as burlap), removing it in the daytime.

- **Target your heating:** Rather than heating the whole greenhouse, use a heated propagator or heat mat to protect specific plants, such as early-sown but tender tomato and chilli seedlings.

- **Power with renewables:** Electric heaters are better than paraffin and propane gas heaters, as long as you're with a renewable energy supplier. Or generate your own electricity – small solar panels used for powering motorhomes can heat greenhouses overnight too. Photovoltaic glass – glass-like solar panels – are still in development, but promise one day to help our greenhouses heat themselves.

- **Run your heater efficiently:** Monitor temperatures with a min./max. thermometer and switch the heater on only when the temperature drops below 7°C (45°F). Plug your heater in via a separate thermostat.

- **Trap the heat in:** Avoid plastic bubble wrap – thick cardboard does the same insulating job. Line the walls and, whenever it gets really cold, line the roof at night, slotting the cardboard sheets behind string netting. Remove the cardboard sheets during the day.

MAKE YOUR OWN GREENHOUSE HEATER

If you've ever sat on a rock in summer and burned the back of your legs on the heat it radiates, you'll know all about passive solar energy. We can collect, store, and distribute this energy to heat our buildings.

Some materials are really good at absorbing and holding on to the sun's energy. Water is the best – absorbing about twice as much as stone, brick, or concrete. Anything painted black stores more heat too. Once the sun has gone down, these materials release the warmth back into the atmosphere.

PASSIVE SOLAR USED TO HEAT HOUSES CAN STORE UP TO **75 PER CENT** OF THE SUNLIGHT THAT COMES THROUGH THE GLASS AS **ENERGY.**

STORING NATURAL HEAT

Sustainable building designers are increasingly using passive solar to heat houses, storing up to 75 per cent of the sunlight that comes through glass as energy. In greenhouses, too, passive solar energy can help keep your plants largely frost free. Deck out your greenhouse with built-in passive solar designs and you can cut heating requirements to almost nothing.

DESIGNS TO KEEP THE HEAT IN

Lean-to greenhouses, or wall greenhouses, are attached to one wall of your house, essentially forming a giant solar heater for your greenhouse. Choose a sunny location and they will draw in heat all day, then gently release it again after the sun goes down. Sunken greenhouses are buried with their lower half about 1.5–2m (5–6½ft) below ground and lined with blocks, bricks, or stone to hold on to heat for longer.

PASSIVE HEATING TRICKS

You don't need a new greenhouse design to start using passive solar systems – you can add these features to an existing structure too. All it takes is a few household objects and a little ingenuity.

- **Plastic bottles:** Paint plastic drinks bottles black and fill with water for mini-heat sinks, or fill jugs of water and dot around the greenhouse.
- **Barrel walls:** Stack black barrels in a wall along one side of the greenhouse to create instant thermal mass. You'll need 20 litres (5 gallons) of water for every 30sq cm (4½sq in) of glass.
- **Install a pond:** If you have space in your greenhouse, a half-barrel pond sunk into the ground doubles up as heat sink in winter and air humidifier in summer.

A HOME-MADE GREENHOUSE HEATER

The problem with installing temporary passive solar is that it takes up valuable growing room. But if you bury it in the ground and add a pump to circulate warm air around the greenhouse, it takes up hardly any space at all.

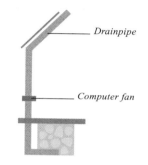

Drainpipe

Computer fan

2 *Fix a second-hand computer fan to the pipe, with a length of drainpipe leading out the other side into the body of the greenhouse. This sucks warm air from the greenhouse during the day and sends it through the heat sink, warming it up further.*

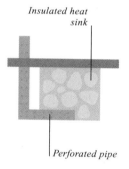

Insulated heat sink

Perforated pipe

1 *Dig a hole 1m (3ft) wide and deep inside your greenhouse and insulate it with old paving slabs or bricks. Feed a length of perforated pipe (with small holes to allow warm air through) into the centre. Then pack around it with stones, bricks, or small recycled plastic bottles filled with water.*

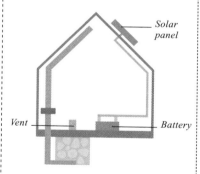

Solar panel

Vent

Battery

3 *Bury another short length of pipe in the top of the heat sink to act as a vent. At night, this releases the warmed air from the heat sink, keeping the greenhouse frost free. Connect the computer fan to a 10-watt solar panel on the roof via a rechargeable 12-volt battery and switch on!*

GREENHOUSE GROWING WITHOUT THE GREENHOUSE

You can significantly reduce the carbon footprint of your greenhouse by using greener heating options (see pp.64–67), but you can't get rid of its carbon footprint altogether. The fact is that a greenhouse is an artificial growing environment that brings with it a large carbon cost.

It's a painful question, perhaps, given how much we all covet one, but are you sure you really need a greenhouse? Unless you're a specialist grower of, say, orchids, the answer is probably no. There is very little you can achieve with a greenhouse that you can't do almost as well without, and with much lower carbon emissions.

FOOD PRODUCTION ACCOUNTS FOR AROUND **25 PER CENT** OF **GLOBAL** GREENHOUSE GAS **EMISSIONS**.

NO GREENHOUSE REQUIRED

There are ways to protect your plants from the weather without a greenhouse. You'll save space in your garden and reduce your carbon footprint.

- **Sheltering frost-tender plants:** Place tender plants in a small, closed cold frame, lined with cardboard for insulation. Water plants only when they are almost bone dry, and on frosty nights, cover the frame overnight with old carpet.
- **Raising seedlings:** Use a heated propagator inside the house with LED grow lights set to give 10–12 hours of daylight. This should germinate all kinds of seeds, from tomatoes in February to climbing beans in April (see pp.136–37).
- **Tomatoes:** There are now several tomato varieties that are resistant to late blight (a serious fungal disease causing plants to turn brown and die). These will grow happily outside, especially given our longer, hotter summers nowadays. Try 'Mountain Magic', 'Crimson Crush', and 'Oh Happy Days'.
- **Melons:** It is now possible to ripen melons outdoors. Go for early-ripening cultivars such as melon 'Emir', and cover with cloches (glass plant coverings).
- **Winter salads:** You can grow winter varieties of lettuce and hardy salad greens outside under cloches, even through the coldest months. Suitable species include: mizuna and mustard 'Red Frill'. Annual herbs, such as moss curled parsley and chervil, will also grow well under cloches in colder periods.

"There is very little you can achieve with a greenhouse that you can't do almost as well without."

GROWING VEGETABLES

Food production, processing, and transportation are responsible for one-quarter of the world's greenhouse gas emissions – so if you want to fight climate change, grow some vegetables.

VEGETABLES THAT DON'T COST THE EARTH

Just over a quarter of global greenhouse gas emissions are generated by the food we eat. Home-grown produce has a much lower carbon footprint than that produced via conventional agriculture. So, start growing your own veg to cut your carbon footprint.

Meat, dairy, eggs, and seafood account for just under a third of the greenhouse gas emissions from food production, but more than a fifth – 21 per cent – comes from growing cereals and vegetables for human consumption. Food grown at home,

---

PLOUGHING EMITS **3 TONNES (3.3 TONS)** OF **CARBON** PER **HECTARE (2.5 ACRES)**, AND IS RESPONSIBLE FOR UP TO **20 PER CENT** OF THE CARBON IN OUR **ATMOSPHERE.**

on the other hand, doesn't need transporting, packaging, or washing with chlorine, and you pick just before you eat, so it rarely makes it as far as a refrigerator, greatly reducing its carbon footprint.

How you grow your veg can make a big difference in how much carbon you save, though. Traditional kitchen gardening techniques often mimic non-organic, commercial farming, using artificial fertilizers, pesticides and herbicides, and ploughing – better known to veg growers as the back-breaking chore of double digging. All carry with them sky-high carbon emissions.

But your garden doesn't have to produce food on an industrial scale, and you're free to do things differently. So, take it easy; garden with a light touch – in tandem with nature – and let your plants thrive.

DON'T DIG FOR VICTORY

Putting away your spade makes a real difference to the carbon cost of the food you grow. Turning over the soil degrades the topsoil and oxidizes the carbon it contains, so most of it floats into the atmosphere as carbon dioxide. Stop digging and add mulches to the soil's surface instead and you leave its structure intact, building over time into carbon-rich seams of organic matter.

Your plants can plunge their roots into an undisturbed subterranean ecosystem of interconnected relationships, from tiny bacteria that help plants capture nitrogen from the air, to mycorrhizal fungi that link roots with soil nutrients.

On a garden scale, the spades-off approach can produce rich rewards. Long-running trials show a no-dig 7.5sq m (81sq ft) bed yields an extra 10kg (22lbs) of veg each year.

ESTABLISHING A NO-DIG VEGETABLE BED

1 *Mow turf close and cut down weeds for the area of your veg bed. Don't dig out turf or weeds, even perennial ones (see p.98). Roots will only grow back if you try to dig them out.*

2 *Lay thick cardboard over the top. As it rots, it will allow vegetables to grow through but will smother annual weeds and weaken perennial ones. If they grow back, add more cardboard.*

3 *Add sides to the bed, either temporary, used to keep the mulch in place while the bed is establishing, or permanent raised beds made of planks or scaffolding boards nailed to pegs driven into the ground at each corner.*

4 *Mulch with organic matter to a depth of 10–15cm (4–6in). Home-made compost, composted bark, or recycled green waste work well, but avoid farmyard and stable manure as it can be contaminated with the herbicide aminopyralid.*

5 *Plant straight into the mulch after treading it firm. Drop seedlings into individual holes or sow in shallow drills in the mulch. Weeding should be minimal, as you've now safely buried weed seeds; just top up with a 2.5–5cm (1–2in) layer of mulch each autumn.*

FOREVER FOOD

Vegetable growers are unique among gardeners in sowing almost everything from scratch each year. Perennial food gardens, though, are considerably less effort than a conventional vegetable patch.

It's a monumental amount of effort to sow vegetables every year. Imagine planting an ornamental garden with nothing but annual flowers each season. But we've become accustomed to a diet of almost exclusively annual vegetables, from peas and potatoes to sprouts and spinach, bred to grow fast, produce a harvest, then die.

Annuals are the least climate friendly of the crops: they demand that you repeatedly disturb the soil by sowing, planting, and clearing, releasing the locked-up carbon. And the faster the crop grows, the higher its carbon emissions: it needs more fertilizers and more watering. Harvesting, say, lettuces removes their entire biomass, continuously depleting your soil's reserves.

Vegetables such as broccoli, peas, or courgettes are better, because stems and leaves return to the soil as compost. Grassy crops, too, such as quinoa and sweetcorn, have fibrous roots that sequester lots of carbon – leave them in the ground after harvesting for your soil to benefit.

Perennial food plants, though, grow year after year, sequestering and storing ever more carbon with no replanting required, so the soil stays undisturbed. You may grow a few already: if you have blackcurrants, apples, or Mediterranean herbs such as rosemary, part of your diet already comes from a perennial food source.

FOREST GARDENS

Perennial food gardens, often known as forest gardens, are more like ornamental gardens to care for. They combine fruit trees with shrubs, such as gooseberries or blackcurrants, and shade-loving leafy greens.

Keep a sunny area clear for perennial vegetables. Some are familiar: you've probably enjoyed spring asparagus shoots or globe artichokes. But many are less commonly grown, so try them on a small scale first to see which you enjoy eating. You can still include annuals – squash, perhaps – but as highlights, much as you would use annual flowers in an ornamental garden.

"Perennial food plants grow year after year, sequestering and storing ever more carbon with no replanting required."

TEN VEG THAT COME BACK EVERY YEAR:

All these vegetables have a hardiness rating of H6 (−20 to −15°C/−4 to 5°F).

- **Babington's leek** (*Allium ampeloprasum* var. *babingtonii*): It resprouts when cut to just above soil level, and dies back naturally in summer before returning in autumn.
- **Caucasian spinach** (*Hablitzia tamnoides*): A shade-tolerant, lush climber, it has pleasantly flavoured young leaves, like spinach.

- -

GROWING **CLIMATE-FRIENDLY PERENNIAL** VEGETABLES SAVES **TIME** AND **MONEY.**

- -

- **Cinnamon vine** (*Apios americana*): This is a pretty climber that produces small, floury tubers like potatoes with a hint of bean.
- **Shieldwort** (*Peltaria alliacea*): This low-growing and shade-tolerant salad leaf has a spicy, garlicky flavour.

- **Chinese artichoke** (*Stachys affinis*): A vigorous mint relative, it produces creamy tubers with a crisp, slightly lemony flavour.
- **Broccoli 'Nine Star Perennial'** (*Brassica oleracea* 'Nine Star'): Produces white cauliflower-like sprigs, produced for up to five years if it's prevented from flowering.
- **Ornamental cabbage** (*Brassica oleracea* varieties): Choose from several varieties, including tree kale, which grows to more than 2m (6½ft) tall, to the more manageably sized Daubenton's kale.
- **Salsify** (*Tragopogon porrifolius*): The clumps of salsify's slender roots have a delicate flavour; leave some in the ground to grow on when you harvest.
- **Sea kale** (*Crambe maritima*): This is a hefty plant with glaucous leaves; new shoots are particularly delicious when forced in spring.
- **Crummock** (*Sium sisarum*): Similar to salsify, crummock tastes like sweet potatoes. Parboil then rub the skins off rather than peeling.

SAVING FOR A RAINY DAY

Though you've picked everything by the end of a long season, don't stop yet: there's one more harvest to gather. Let a few fruits ripen, collect their seed, and you have everything you need to sow next year's crops.

Seeds are the stock in trade of any vegetable gardener raising annual crops. But commercial seed production is an international business. The seed you buy may have been bred in one country, tested in another, and produced in a third before it's finally shipped back to be sold to you – with

---

GLOBAL **GREENHOUSE GAS** EMISSIONS, SUCH AS **METHANE** AND **NITROUS OXIDE**, DUE TO **FOOD PRODUCTION** ARE A SIGNIFICANT **DRIVER** OF **CLIMATE** CHANGE.

---

all the extra carbon emissions that involves. So, source seed locally and check the seed company grows and harvests its own seed.

Save seed from your own plants, though, and they're as local and low carbon as it gets. Seeds from plants in your garden are already adapted to perform well in your soil and climate, so you should see better results with each generation – in effect, you're breeding your own heirloom varieties.

Only save seed from open-pollinated plant varieties, so you can be sure the offspring will be the same as the parents. Select your biggest, strongest plants with the genetic qualities you want for their offspring. Tie a piece of brightly coloured wool round the stem so you remember not to pick from them.

Some plants, such as peas, beans, and tomatoes, are self-fertile. For cross-pollinating vegetables, such as lettuces and brassicas, let just one variety flower and set seed each year to keep them true to their type.

EASY SEEDS TO SAVE

- **Beans and peas:** Once pods turn yellow, harvest and shuck out the seeds. Spread out on a tray to dry for a fortnight before storing in a paper bag or envelope.
- **Annual herbs:** Put a paper bag over ripe seedheads, cut the whole head, and upend it, shaking to loosen the seeds. Pick out chaff and debris, then store cleaned seed in a paper bag or envelope.
- **Tomatoes:** Cut a ripe fruit in half, scoop out seeds and pulp, spread onto kitchen towel, and leave to dry on a windowsill. Store, kitchen towel and all, in a paper bag or envelope. Next spring, lay the kitchen towel on a tray of compost, cover with more compost; the seeds should germinate.

HOW TO HAND POLLINATE

Some vegetables, such as squash and pumpkins, cross-pollinate easily, resulting in smaller, less tasty, and misshapen fruits. Keep varieties "true" by hand pollinating individual flowers.

Female flowerbud

Fruit

Male flower

1 *Locate a female flowerbud, which has a swelling fruit below the petals, and several male flowers. Male flowers have a straight stem.*

Female flowerbud

2 *When the buds flush yellow, they're about to open the following day: gently tie the female bud shut with wool or use a clothes peg.*

Anthers

3 *Next morning, pick off the male flowers and remove all the petals to expose the pollen-covered anthers.*

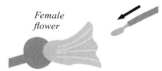

Female flower

4 *Untie the female flower, then dab the exposed anthers into the centre, repeating with other male flowers to ensure genetic diversity.*

5 *Tie or peg the female flower shut again and tie a label to the stem so you remember which fruit to save seeds from.*

SEEDY JARGON BUSTER

F1 seed: *"first generation" seed produced artificially by cross-breeding the same two parents; note: offspring from saved seed won't be the same.*
Open-pollinated seed: *naturally pollinated, then selected over several generations; saved seed will mostly be just like the parent variety.*

Self-fertile: *flowers on the same plant fertilize each other, so the offspring is the same variety, too.*
Cross-pollination: *when two separate plants pollinate one another; if they are the same variety, the offspring is "true" but if parents are different the resulting hybrid has traits from each.*

WATER GARDENING

Bodies of water can absorb carbon dioxide from the atmosphere. So, our garden ponds are not only fascinating habitats for plants and wildlife, but can also act as carbon sinks.

POND LIFE

You can lose yourself for hours in the watery world of a garden pond. Settle down close to the water and watch the dramas unfold: a flashing dragonfly on the hunt, tadpoles magically metamorphosing, and zig-zagging pond skaters – proof you can walk on water if you're small enough.

Ponds make fascinating features for your garden, but they are also amazingly efficient at absorbing greenhouse gases. Every part of a pond captures carbon dioxide, including plants, water, and sediment. In fact, natural ponds can absorb up to 20 to 30 times as much carbon as forests or grassland.

Unfortunately, the world's most efficient carbon sinks can also be among the worst emitters of methane, a potent greenhouse gas. One Swedish study found just 12 of 40 ponds analyzed acted as carbon sinks: 28 emitted more greenhouse gases than they absorbed.

The methane comes mainly from the muddy gloop on the pond floor, mostly dead plant material rotting in anaerobic conditions. Ponds are shallow, so there's no time for the water to absorb the methane before it bubbles to the surface.

The problem is worse in stagnant ponds or those that dry out, exposing sediment directly to the air. High nutrient levels and warming summer temperatures all tip the balance towards a pond that contributes to global warming instead of offsetting it.

There's little analysis of the climate impact of garden ponds in particular. The obvious distinction between garden ponds and natural ponds, though, is that garden ponds are gardened. So, you, the gardener, can be the difference that transforms your garden pond into the efficient carbon sink it was designed to be.

TWO RECENT **AUSTRALIAN** STUDIES SUGGEST THAT **FARM PONDS** AND **URBAN LAKES** EMIT MORE **METHANE** THAN **LARGER** WATER BODIES IN THE SAME REGION.

"The world's most efficient carbon sinks can also be among the worst emitters of methane."

MANAGING A LOW-CARBON POND

- **Design your pond to attract wildlife:** All creatures visiting your pond add biomass and therefore carbon to your garden. Plus, watching wildlife in ponds is a very low-carbon activity.
- **Line your pond with clay:** Compacted or "puddled" clay makes a watertight liner and may help your pond absorb more carbon as it's a natural material.

LEAVE **SEDIMENT UNDISTURBED** TO **REDUCE METHANE** EMISSIONS.

- **Plant lots of pond plants:** Aim to cover two-thirds of the pond's surface, including marginals (plants that grow at the shallow edges of a pond), such as marsh marigolds, and deep-water aquatics, such as waterlilies.

- **Clear spent foliage promptly:** Snip away fading flowers in summer and remove dying foliage in autumn before they decay into the water.
- **Net ponds in autumn:** Falling leaves cause silt to build up, so site ponds away from trees, fish out floating leaves promptly, and use nets to prevent leaves falling in.
- **Don't use fertilizers near ponds:** Lawn feeds and other nitrogen-rich fertilizers leach into the water and bump up nutrient levels, turning ponds stagnant.
- **Keep your pond topped up:** When ponds dry out, especially if sediment is exposed, they emit methane directly into the air.
- **Top up with rainwater:** Tap water contains nutrients that help to turn ponds stagnant, so use saved rainwater if you have it.
- **Leave the sediment undisturbed:** Every handful you bring to the surface releases methane, so avoid clearing out your pond. If you reduce dead vegetation in and around the pond, you shouldn't need to.

MAKE A
WILDLIFE POND

Natural ponds are vanishing fast. Half a million ponds have disappeared from the UK alone in the last century. So, the ponds in our gardens offer a last haven for the two-thirds of freshwater species that depend on them.

Ponds provide habitat and food for toads, frogs, and diving beetles, as well as mayflies, damselflies, and newts. They also act as service stations for larger creatures such as birds, bees, and bats in need of a drink. They'll sort out your pests,

---

EVERY **CREATURE** THAT VISITS YOUR **POND** ADDS **BIOMASS** TO THE **CARBON-SINK** GARDEN.

---

hoovering up caterpillars and slug eggs so that your plants stay healthy without pesticides. And every creature that arrives adds biomass to your carbon-sink garden.

One of the delights of installing a garden pond is how quickly it attracts wildlife: it seems you've only just finished filling it when the first visitors arrive. They'll turn up for the tiniest patch of water, as small as a washing-up bowl, though the larger your pond, the more species it will attract.

Ideally, aim for an area that is at least 1 by 2m (3 by 6½ ft). Choose a sunny spot with some afternoon shade, and avoid nearby trees that could drop leaves into the water.

Shallow areas warm up quickly and are popular egg-laying spots; deeper sections are cooler and offer shelter from predators. Allow six to nine plants per sq m (11 sq ft) – or just one large plant like a waterlily. Pond dwellers like different plant types, so cater for as many species as you can.

PLANTS FOR YOUR POND

- **Oxygenators:** Submerged plants, such as hornwort (*Ceratophyllum demersum*), water crowfoot (*Ranunculus aquatilis*), and common water starwort (*Callitriche stagnalis*), keep the water clean by producing oxygen and taking in nutrients.
- **Floating plants:** Floating plants provide shelter and keep the pond cool. Try waterlily (*Nymphaea alba*), frogbit (*Hydrocharis morsus-ranae*), and water crowfoot (*Ranunculus aquatilis*).
- **Emergent plants:** Also known as marginals, these grow on the pond edge, with their roots in the soil underwater and foliage above water. Try dwarf bulrush (*Typha minima*), pond sedge (*Carex riparia*), and yellow iris (*Iris pseudacorus*).

CREATE A HAVEN FOR WILDLIFE IN YOUR POND

As you dig the hole for your pond, create gentle contours with different planting depths, avoiding vertical "walls", down to a deepest point of 60–90cm (2–3ft) where pond dwellers can overwinter safely.

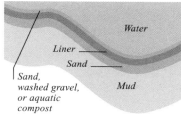

Water

Liner ———

Sand ———

Sand, washed gravel, or aquatic compost

Mud

1 *Spread a thick layer of sand over the bottom of the hole to protect the liner and place the liner on top. Sand, washed gravel, or aquatic compost on top provides habitat for wildlife.*

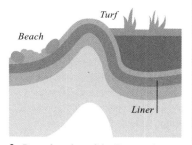

Turf

Beach

Liner

3 *Bury the edge of the liner under rocks or turf to blend the edges of the pond more naturally with the garden.*

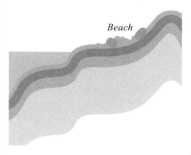

Beach

2 *Make a sloping beach at one end, lined with areas of children's playpit sand, washed gravel, and pebbles to create lots of opportunities for egg laying as well as access for larger creatures.*

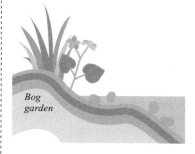

Bog garden

4 *Include a bog garden alongside the pond using marginal plants such as marsh marigold* (Caltha palustris) *and purple loosestrife* (Lythrum salicaria) *to act as a green corridor between the pond and the wider garden.*

CREATE A RAIN GARDEN

Rain gardens are shallow dips in the ground that absorb excess rainwater then release it slowly once storms subside. This percolation into the ground (instead of running straight off the surface), also reduces flooding, protecting nearby land and buildings.

As well as protecting your house from flooding, a rain garden is as good as a pond for absorbing carbon. Often, it is better, as rain gardens tend to be densely planted with high-biomass plants such as ornamental grasses. Their roots are particularly good at storing carbon beneath the soil.

There's no pond liner between the water and soil, so all that carbon-rich humus can sink directly into the ground to be stored. In one study, rain gardens captured and stored more carbon than any other storm-water-retention scheme.

Rain gardens attract lots of wildlife because they are effectively seasonal ponds. Such ponds dry out in summer and refill in winter, creating an ever-changing landscape with muddy spots where dragonflies lay their eggs and warm shallows where frogs spawn.

All you need to make one is a spade, some pebbles, a downpipe from a nearby roof and of course plenty of plants – so your carbon footprint is pretty low, too!

RAIN CATCHER

A simple rain garden will absorb about 30 per cent more rainfall than a lawn.

Plants for a rain garden must be able to cope with waterlogging as well as drought.

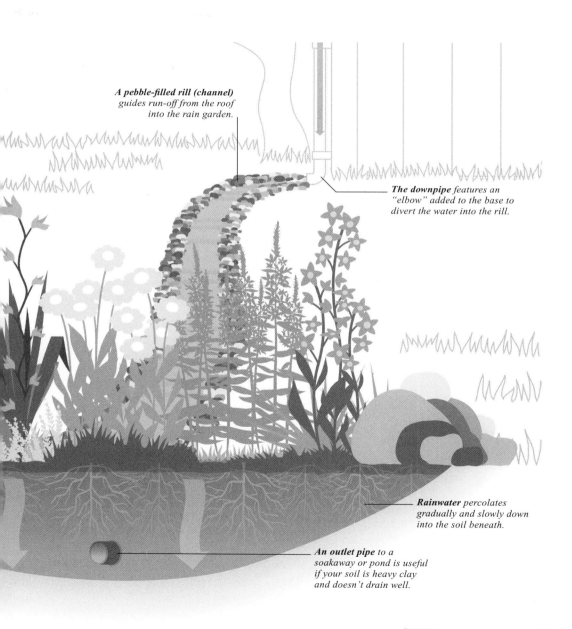

A pebble-filled rill (channel) guides run-off from the roof into the rain garden.

The downpipe features an "elbow" added to the base to divert the water into the rill.

Rainwater percolates gradually and slowly down into the soil beneath.

An outlet pipe to a soakaway or pond is useful if your soil is heavy clay and doesn't drain well.

WATERING

Every time you open the tap to water your garden with treated water, you add to your carbon footprint. So, use water with care and make the most of what falls from the sky – it's free, and it's better for your plants and the environment, too.

THE LOW-WATER GARDEN

Watering the garden is one of those traditional garden chores, up there with weeding, digging, and mowing the lawn in the list of things you "ought" to do to be a good gardener. But perhaps it's time to rethink the way we've always done things.

The water we use for our gardens, especially in summer, is often treated tap water, frequently drawn from natural wetlands. As they're drained to supply reservoirs, one of our most efficient carbon sinks turns into a net contributor to climate change as methane-emitting sediments are exposed to the air.

Every litre, or gallon, of water you use adds to your carbon footprint. Rainwater is better than treated tap water, but storing it in water butts and tanks still entails the manufacture, transport, and installation, usually of

GLOBALLY, **WATER SUPPLIES** ARE UNDER INCREASING **PRESSURE** FROM THE EFFECTS OF **CLIMATE CHANGE** AND RISING **DEMAND** FROM THE WORLD'S **GROWING** POPULATION.

plastic – and therefore pollution-heavy – goods. A truly low-carbon garden is one that doesn't need much watering at all.

HOW TO TURN OFF THE TAP

- **Use a watering can, not a hose:** Hoses can get through 1,000 litres (265 gallons) of water an hour – two days' supply for a family of four in the UK. Using watering cans targets water at the roots and slows the flow, soaking the compost more thoroughly.
- **Mulch, mulch, mulch:** Keep soil covered with a 5cm (2in) layer of organic matter to prevent water evaporating and increase the soil's ability to hold on to moisture.
- **Grow perennials, not annuals:** Annual flowers demand daily watering for their first few months, while they're seedlings and young plants. Perennials, once established, should need no water at all.
- **Cultivate drought-tolerant vegetables:** Growing food is thirsty work, so major on veg that

> *"Every litre, or gallon, of water you use adds to your carbon footprint."*

can survive without much water, such as globe artichokes, chard, leeks, beetroot, and other root veg.

- **Grow in the open ground:** Plants can send roots down a long way to find moisture, so avoid planting in containers, if you can, or planting in greenhouses.
- **Plant in autumn, not spring:** In autumn the ground is warm and damp, so plants develop a good root system naturally, without needing extra watering.
- **Plant the right plant in the right place:** Match plants with the spot they like best, whether that's sunny, shady, free draining or water retentive, and the natural water level will suit them too.

ZERO-WATER PLANTING

Xeriscaped features, in which plants are suited to very dry landscapes such as gravel or scree gardens, will never need watering. The RHS Garden in Hyde Hall in Essex, where the average annual rainfall is lower than in Beirut, Lebanon, hasn't been watered since it was created in 2001. That said, you'll lose all your plants in the first wet winter if you live somewhere with higher rainfall, or garden on moisture-retentive clay soil. For zero-water gardens in a climate-warmed world of extreme winter rainfall, you need plants that can cope with both drought and flood equally well. Choose from among these species:

- **Trees:** Himalayan birch (*Betula utilis* subsp. *jacquemontii*), bull bay (*Magnolia grandiflora*), and crab apples (*Malus sylvestris*).
- **Shrubs:** Rose (*Rosa rugosa*), Siberian dogwood (*Cornus alba* 'Sibirica'*)*, and hydrangea (*Hydrangea arborescens*).
- **Perennials:** Geranium, geum, and daylily (*Hemerocallis)* species.
- **Grasses:** Korean feather reed grass (*Calamagrostis brachytricha*) and eulalia (*Miscanthus sinensis*).

CATCH THE RAIN

Even when you've cut watering to a minimum, there will be times when you reach for a watering can, especially as we learn to live with the hotter, drier summers of a climate-changed world.

Relying on treated tap water is the high-carbon option. Treating water has a relatively low carbon footprint, but it takes huge amounts of energy and infrastructure to pump it between reservoirs, treatment centres, and your home.

ABOUT 24,000 LITRES (6,340 GALLONS) OF **RAINWATER** CAN BE CAPTURED FROM AN **AVERAGE** HOUSE **ROOF** EACH YEAR – EASILY **ENOUGH** TO COVER YOUR **NEEDS**.

Climate change is also bringing us warmer, wetter winters, though, dumping rain in floodwater quantities on our homes and gardens. Catch this bounty and save it for summer droughts, and you'll reduce stormwater run-off and pollution and even out the rollercoaster ride.

Plants love rainwater. It's loaded with dissolved nitrogen and is largely chemical free. Rainwater can – rarely – carry water-borne diseases such as damping-off disease, so you

might choose tap water over rainwater if you're watering seedlings, say.

About 24,000 litres (6,340 gallons) of rainwater can be captured from an average house roof each year – easily enough to cover your needs if you take the steps outlined on pp.88–89 to minimize the amount of water your garden requires. All you have to do is find somewhere to put it all.

WATER-STORING SOLUTIONS

- **Rain saucers** catch rainwater without water butts, via an inverted "umbrella" of strong plastic, like old compost sacks. Feed via a pipe into any free-standing container.
- **Water butts** come in all shapes and sizes; wooden barrels and galvanized cattle troughs make good plastic-free alternatives. Link several together to maximize capacity.
- **Dipping ponds** are fed direct from downpipes and double up as wildlife attractors. Raise the sides to keep soil out and water clean.
- **Above- and below-ground tanks** give you up to 10,000 litres (2,640 gallons) of storage capacity. You'll need plenty of room, and the carbon emissions from manufacturing and installing them can be considerable.
- **Blue roofs** save water in a tank on the roof. Consult a specialist structural engineer as stored water is extremely heavy.

HOW TO LINK TWO OR MORE WATER BUTTS

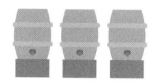

1 *Place your water butts side by side in a row, raised on breeze blocks or bricks to allow a watering can underneath each tap.*

2 *Link the downpipe to the first water butt in the normal way, adding a filter to catch debris.*

Hole

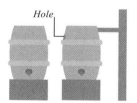

3 *Now make a hole in the water butt opposite the downpipe, about a third of the way from the top. Avoid linking butts at the base as the connecting pipe will get blocked by debris.*

Hole

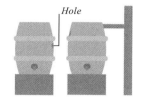

4 *Make a corresponding hole in the second water butt, lower than the first so water flows easily from one butt in to the next.*

5 *Link the two butts with a short length of hose, held in place with glue and silicon sealant, or buy a water butt connecting kit.*

6 *Repeat the process to add more water butts.*

HARVEST YOUR RAINWATER

Usually we use drinking-quality water (which has been treated), but really there is no need. Save enough rainwater and you can use it for all that your garden and plants require.

Install a rainwater harvesting system and you can offset the carbon emissions from watering not just your garden, but your entire home.

There are two designs: underground and above ground. Underground tanks are buried and use a pump to get water to your house; these require major earthworks to install but afterwards are almost invisible. Above-ground tanks are usually smaller but you can link several together; they're easy to install, but take up space and can be unsightly.

A typical 3,000-litre (660-gallon) tank holds about five days' supply for a family of four. In winter, that's enough for your needs, but in summer you will usually need to switch back to mains water, especially in droughts.

To keep rainwater separate from drinking-quality water, you will need two sets of plumbing, so harvesting systems are easier to install in new-build homes or when renovating. If you are retro-fitting them to an existing building, above-ground systems are a much more straightforward choice.

HARVESTING SYSTEMS

Such systems collect, store, and redistribute the rainwater that falls on roofs of sheds and houses.

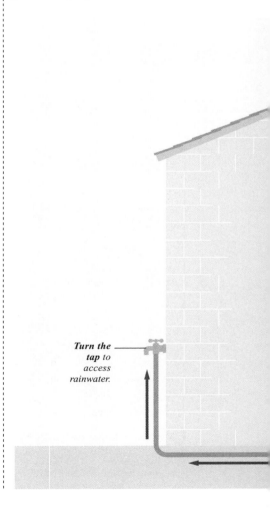

Turn the tap *to access rainwater.*

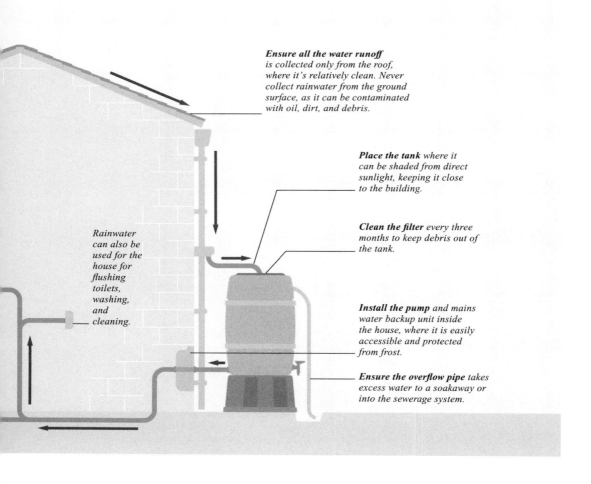

Ensure all the water runoff
is collected only from the roof,
where it's relatively clean. Never
collect rainwater from the ground
surface, as it can be contaminated
with oil, dirt, and debris.

Place the tank where it
can be shaded from direct
sunlight, keeping it close
to the building.

Clean the filter every three
months to keep debris out of
the tank.

*Rainwater
can also be
used for the
house for
flushing
toilets,
washing,
and
cleaning.*

Install the pump and mains
water backup unit inside
the house, where it is easily
accessible and protected
from frost.

Ensure the overflow pipe takes
excess water to a soakaway or
into the sewerage system.

WEEDING

Weeds have a bad reputation, but perhaps it's time to change the way we think about them. We can manage our weeds and welcome the wildflowers, without racking up our carbon footprint.

HOW TO DEAL
WITH THE WEEDS

Weeds are pretty extraordinary plants, when you think about it. They grow faster and taller, and spread more persistently than any of their neighbours, and they've evolved some impressive survival strategies.

Weeds can regenerate from roots many metres (feet) deep in the soil, use other plants' stems to climb up to the light, and fire seed in all directions to make sure you won't find them all, however hard you try. Indeed, weeds are so successful that they'll outcompete garden plants for water, food, light, and growing space, and most gardeners spend a lot of time fighting back.

Keeping your garden well stocked and mulched will help to elbow out the weeds. Weeds are really just nature's way of recolonizing bare

earth, so if you keep your ground well covered with a "living mulch" of dense planting, you'll push out the weeds, too.

Where weeds do get a foothold, you can stage an effective counterattack without adding to your carbon footprint simply by hand weeding rather than reaching for a spray bottle. And if you take a low-carbon approach to hand weeding, that's less of a chore than it sounds.

For a start, you can keep your garden fork stashed away in the tool bag. The more you dig weeds out by the roots, the more of the soil's carbon you turn to the surface to oxidize. Besides, you also expose new weed seeds to germinate, so it's a pretty pointless exercise anyway. Here's what you can do instead.

BURY WEED SEEDS

The top 45cm (18in) of your soil is crammed with weed seeds waiting for the right conditions to germinate: sometimes, if they're goosefoot or poppies, for up to 40 years. Often, all

THE **PRESENCE** OF CERTAIN **WEEDS** CAN INDICATE A **DEFICIENCY** OR **EXCESS** OF SOIL **NUTRIENTS**, SUCH AS **CALCIUM**, **NITROGEN** OR **PHOSPHORUS**.

> *"The top 45cm (18in) of your soil is crammed with weed seeds waiting for the right conditions to germinate."*

they need is light, usefully provided by the gardener lifting them to the surface by turning the soil.

Instead, bury them out of harm's way under a 5cm- (2in)-thick mulch of organic matter. Composted green waste from your local council is sterile and won't introduce more weed seeds like manure and garden compost can.

COMPOSTED **GREEN** WASTE FROM THE COUNCIL IS **STERILE** AND WON'T INTRODUCE MORE **WEEDS**.

HOE, HOE, HOE

You'll reduce your annual weed-seed problem almost to nothing by mulching, but you can't prevent the blow-ins: seeds that arrive on the wind, deposited in bird faeces, on the soles of your boots, or via the paws of your dog or cat.

Hoeing beheads weed seedlings while they're still tiny. Choose a dry day, so severed weeds wither on the surface, and gently sweep the blade to and fro just underneath the mulch. Repeat fortnightly through the growing season.

FINGER WEEDING

When weeds root into mulch, they're only just clinging on in a light, crumbly surface. So when you're weeding close to plants and a hoe is too clumsy, or if you only have a small area to deal with, your fingers are all you need.

Pull up individual weeds between thumb and finger, or just use your fingertips like miniature hoes. Ruffle up the surface of the mulch to break it up and you'll upend any rooting weeds, ready to collect and cart off to the compost bin.

KILLING WEEDS
WITHOUT THE WEEDKILLER

Every gardener knows that persistent weeds are an unfortunate certainty. Many people reach for the spray bottle of weedkiller, but there are many benefits to shunning artificial chemicals.

Persistent perennial weeds such as bindweed, ground elder, couch grass, and mare's tail return year after year, seemingly whatever you do. You might think you dug every last bit out by the roots, but like a plant zombie apocalypse, they rise again to shoulder aside your much-loved garden plants and swamp your carefully tended crops. It's at this point that many gardeners resort to the weedkiller.

You're right not to simply dig weeds out – digging releases carbon and makes the problem worse by breaking underground roots, each broken end producing a new plant. But turning to herbicides creates a new raft of problems.

HARMFUL EFFECTS

The most widely used weedkiller is glyphosate, though there are other products, such as triclopyr for killing stumps and woody weeds, and clopyralid, a broadleaf weedkiller for lawns. This last, incidentally, is very persistent and will stunt any plants

you grow in compost made from treated grass clippings.

There's scant information about the carbon cost of manufacturing and using weedkillers, but the persistent concerns about their impact on human health should be enough to stay your hand. Besides, you can kill off most perennial weeds without them – you just need to be as persistent as the weeds are.

WEED BUSTING THE CHEMICAL-FREE WAY

- **Hand pulling:** Like all plants, perennial weeds can't survive without leaves. So, pull or hoe top growth as soon as it emerges and repeat weekly to weaken and, after a few years, kill weeds altogether.
- **Boiling water:** Pour the contents of your just-boiled kettle (heated with energy from renewable sources) over weeds in paving to kill them instantly, down to the roots.
- **Electric weeders:** Using energy from renewable sources, burn off top growth from patio weeds with a targeted "thermal shock" of up to 600°C (1,110°F). Repeat weekly.
- **Southern marigold:** (*Tagetes minuta*) This giant 1.8m (6ft) annual emits chemicals from its roots to stop other plants growing nearby, including persistent weeds such as ground elder.

HOW TO TACKLE LARGE WEEDY AREAS

Transform even the weediest beds into an immaculate area ready for planting right away, with no spraying or digging required.

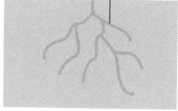

Weed roots can remain

1 *Take off all the top growth to just below soil level, but don't worry about removing the roots – if you try to dig them out they will only return.*

Continuous cardboard layer

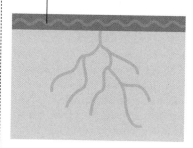

2 *Cover the whole area with a thick sheet of brown cardboard, or two or three layers of thinner cardboard. Overlap sheets so there are no gaps.*

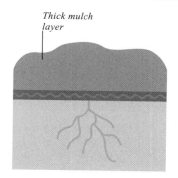

Thick mulch layer

3 *Next, cover the cardboard with a deep layer of mulch, 10–15cm (4–6in) thick, using well-rotted organic material like home-made compost or recycled green waste.*

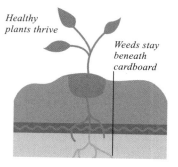

Healthy plants thrive

Weeds stay beneath cardboard

4 *Plant straight into the mulch. Plants will root through the cardboard (you may have to water the area in drier regions), but the weeds underneath struggle to find the light. You may need to repeat the process for a few years until the weeds finally give up.*

LOVE YOUR WEEDS

For many gardeners, weed control means hours of back-breaking work hand-pulling and hoeing weeds. There is one way to get rid of all your weeds without expending any effort – just start calling them wildflowers.

We are so fickle towards plants. We welcome forget-me-not seedlings, but pull out herb robert – an equally pretty plant with its ferny foliage and tiny pink flowers. We go to great trouble to plant delicate wildlings such as lady's bedstraw, cornflowers, and catchfly, hoping they'll attract pollinators, forgetting that the humble dock supports an astonishing 33 different species of moth.

Plants we call weeds can be as pretty as any cultivated perennial. Tall, willowy cow parsley bursts like white froth through borders, and

oxeye daisies seed into gaps with the look of a shasta daisy but without the price tag.

EDIBLE WEEDS

Try growing a "weed" vegetable garden. Pick edible ground elder leaves before they open, and wilt their long, succulent stems in olive oil with garlic and herbs. Blanch dandelion crowns for a mildly bitter, chicory-like salad leaf. Or combine their roots with those of burdock to make a delicious root beer.

NATURE'S HELPERS

Weeds colonize bare ground because they're nature's way of preventing soil erosion. They protect carbon locked into the soil, then add their own payload from carbon-packed roots to nitrogen-rich top growth. Cut them away at ground level, before they seed, compost the tops, and leave roots in the ground to get all these benefits without the drawbacks.

By all means remove weeds you have good reason to dislike, such as bindweed, which strangles nearby plants. And perhaps reserve a spot away from the house for less aesthetically pleasing plants, such as docks and nettles. But where you can, put away the hoe and welcome your wildflowers in to your newly weed-free garden.

WEEDS IMPROVE THE **FERTILITY** OF THE **SOIL**. WHEN WEEDS **DIE**, BOTH THE **LEAVES** THAT FALL TO THE GROUND AND THEIR **ROOTS DECOMPOSE** INTO **FERTILIZER** FOR THE SOIL.

"The humble dock supports 33 species of moth, including beauties such as the ruby tiger and striped hawk moth."

FEEDING

It's time to ditch the synthetic fertilizers – your home and garden may already contain everything you need to keep your plants and carbon-rich soil healthy.

FEED THE SOIL, NOT THE PLANT

The manufacture and transport of synthetic fertilizers has a significant carbon cost. While synthetic fertilizers feed the plants directly, they do little to improve the soil itself. Natural plant foods often enrich your soil with nutrients, too.

In nourished, aerated soil, roots can stretch out and plants thrive.

If you're adding extra synthetic fertilizer to plants growing in the open garden, you're probably doing more harm than good. Most healthy soils already contain all the minerals and trace elements plants need. So adding extra fertilizer to boost growth makes absolutely no difference. And as climate-changing gardening activities go, nitrogen-based fertilizers – especially artificial, non-organic ones – are right up there with petrol mowers, glyphosate, and double-digging.

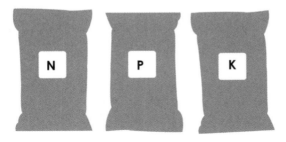

Fertilizers deliver nitrogen (N), phosphorus (P), and potassium (K) in various amounts.

HIDDEN DANGERS

Almost all synthetic fertilizers such as sulphate of ammonia, sulphate of potash, and non-organic tomato feed are created using the Haber-Bosch process. Known as "bread from air", the process pulls nitrogen from the air to make ammonia, an excellent fertilizer. It's an incredibly energy-intensive method powered by vast quantities of fossil fuels.

After you spread synthetic fertilizers, up to 5 per cent of the nitrogen is re-released as nitrous oxide. For every 1kg (2¼lb) of fertilizer you scatter on the garden, you add about 2.7kg (6lb) of greenhouse gases to the atmosphere.

NATURAL NUTRIENTS

Unfortunately, some organic fertilizers are often damaging too. Commercially produced poultry manure pellets are usually stored in anaerobic, methane-belching conditions before processing. Bonemeal, and blood, fish, and bone, have to be pasteurized, sterilized, or

> *"For every 1kg (2¼lb) of fertilizer you scatter on the garden, you add about 2.7kg (6lb) of greenhouse gases to the atmosphere."*

incinerated – all high-energy processes. Seaweed feeds are probably the most climate-friendly; although, they are often wild-harvested, with potentially damaging implications for local marine life, and must be processed and transported to you just like any other commercial fertilizer.

So, all commercially produced fertilizers have some environmental impact. You can do much to keep that to a minimum by using only the right amount and applying it at the right time of year (read the packet information carefully and follow the instructions to the letter).

Adding chemicals might change your soil profile but only temporarily, so you'll need to do it every year. So instead of trying to lower the pH of alkaline soil, it's easier, and certainly more climate-friendly, to grow alkaline-loving plants instead that will thrive in chalk, such as irises or peonies.

So, ditch the packs of fertilizer and focus on keeping the soil healthy and able to do the job it is designed to do.

ZERO-WASTE GARDEN

As we harvest vegetables, prune, weed, and generally tidy up, we're removing leaves and stems that would otherwise fall back onto the soil for earthworms and bacteria to process into soluble, plant-friendly nutrients. All you have to do to keep your garden producing strong, healthy plants is put it back.

So turn as much plant waste as possible into piles of compost (see pp.174–175), and add extras from outside the garden, such as kitchen scraps or manure, to make up the difference. Then spread it back over your beds as a generous 5cm (2in) thick mulch each autumn. That's it: the only "feeding" regime you'll ever need, and it's kind to the planet, too.

Redistribute organic matter back onto your garden.

An autumnal mulch really feeds the soil.

GROW YOUR OWN PLANT FOOD

Synthetic plant fertilizers are carbon intensive and costly to the environment and your pocket. But it doesn't need to be that way: you can find all the natural fertilizer you need at home, and for free.

We all need a pick-me-up from time to time: that reviving cup of coffee that gets you going in the morning, or the cheeky bar of chocolate to perk you up after an exhausting day. And it's the same with plants. Fast-acting liquid feeds added to the watering encourage strong growth, flowers, and fruits.

There's no need to resort to carbon-intensive synthetic feeds: you don't even have to buy organic feeds in single-use plastic bottles. You've probably got most of what you need to hand already, growing in your garden or among the everyday objects lying around at home.

Fermented liquid feeds that break down leaves under water anaerobically, without oxygen, give off a little methane as they brew (that's why they smell). A lower-carbon option is to make aerobic compost tea, bubbling oxygen through the mix. Alternatively, you can just spread your feeds on the surface of the soil to release nutrients slowly and naturally throughout the season.

PLANT FOOD TO MAKE AT HOME

- **Comfrey tea:** Grow non-invasive "Bocking 14" comfrey to make a potassium-rich plant food for better fruiting. Cram a bucket with comfrey leaves, cover with water, and leave for six weeks. Dilute one part with 20 parts water – the colour of weak tea – before using.
- **Nettle tea:** Young nettles are packed with nitrogen. Harvest them at this stage and use on young plants, foliage houseplants, and leafy veg. Lay the leaves on the soil surface to break down slowly, or make a liquid feed as for comfrey (above).
- **Vegetable peelings:** You don't even need a garden to make fertilizer smoothies. Half-fill a blender with fruit and vegetable peelings, top up with water, and whizz into a watery soup (add more water if necessary). Use neat, and as fresh as possible.
- **Urine:** Human urine is very rich in nitrogen, and you'll never run out! Always use fresh (it shouldn't smell), and only if you're not on medication. Dilute 1:50 with water and pour onto the soil, not the plants.

HOW TO MAKE COMPOST TEA

Convert ordinary compost into fast-acting liquid feed by brewing compost tea – a heady brew of microbes and energy for your plants.

Tie the bag with string

1 *Put a spadeful of well-rotted compost inside a cotton or hessian bag and tie the top.*

Let tap water stand for a few hours

2 *Now fill a 20-litre (5-gallon) bucket with rainwater, or use tap water that has stood for a few hours so the chlorine can evaporate. Add five tablespoons of sugar or molasses.*

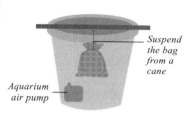

Suspend the bag from a cane

Aquarium air pump

3 *Lower an aquarium air pump into the water and switch it on. Tie the bag to a cane. Balance the cane over the bucket so the bag hangs in the water.*

Transfer liquid feed to a bottle for ease of use

4 *Leave to brew for two days – it will bubble and froth. Once it's done, lift the bag out of the bucket and your tea is ready. Use it neat on your plants and as fresh as possible.*

SEAWEED SUPERFOOD

Seaweed is packed with micronutrients and plant growth hormones: it's phenomenally good as a biostimulant for both plants and soil. Always ask permission from local authorities before taking seaweed from a beach, and pick up beached seaweed rather than pulling it off the rocks (it's an important habitat for coastal wildlife). Spread direct on veg beds in autumn for planting the following spring, add to the compost heap, or make a tea (see the instructions for comfrey feed).

SET UP A WORMERY

Building a wormery, or worm composter, is a fantastic and space-efficient way to reduce your kitchen food waste while also producing nutrient-rich compost and liquid fertilizer to use in your garden.

We are not all blessed with the space in our garden for a conventional compost heap, but there is a solution if you're still keen to produce your own rich compost – set up a wormery. Wormeries take up a lot less space because efficient composting worms do most of the work.

ABOUT 0.5KG (1LB) OF **COMPOSTING WORMS** CAN EAT BETWEEN **0.8KG** (1¾LB) AND **1KG** (2¼LB) OF KITCHEN **WASTE** EACH **WEEK**.

FEEDING YOUR WORMS

Although compost bins are great for recycling raw peelings, it's best to avoid putting cooked kitchen waste (such as meat) in your bin – it rots and attracts flies and rats. You can still turn your leftovers into compost, though – just feed them to the worms.

Worms enjoy a varied diet – some of their favourite kitchen waste includes: fruit and vegetable scraps, cooked pasta, coffee grounds, used tea bags, crushed eggshells, bread, rice, and small amounts of weeds, leaves, and cut flowers.

Wormeries use brandling worms to break down food waste into rich compost, plus worm tea – a superb liquid plant food. You can buy hi-tech wormery kits with separate layers for each stage of composting, but they take energy and resources to manufacture. Besides, it doesn't need to be that complicated: just make your own simple wormery at home (see opposite), with repurposed materials for a fraction of the cost.

LOOKING AFTER YOUR WORMERY

- **Add food** a little at a time, in layers no more than 5cm (2in) thick.
- **Cover the layers** with wetted newspaper or hessian to keep everything damp.
- **Don't include too much citrus** or onion peel as they're very acidic.
- **Avoid meat** and animal products, which can attract vermin.
- **Keep your wormery** in a warm, sheltered, but shady place.
- **Protect your wormery** with blankets in cold weather, or bring it into a garage or shed for winter.
- **Check the wormery tap** for blockages each time you add waste: if the liquid level rises, the worms can drown.

MAKING A WORMERY

Wooden bins (from recycled untreated pallets; see also p.22) make the best wormeries as the air flow is better; you can repurpose plastic bins if you drill plenty of air holes. Aim for a container at least 35cm (14in) across. The lid stops your wormery from getting waterlogged.

Kitchen waste *Add no more than 5cm (2in) at a time. The worms live in the top layer of compost, so once the bin is full, remove the top layer and set aside, dig out the compost to use on the garden, then replace the worms and start again.*

Damp bedding material *You'll need a layer about 10cm (4in) deep of shredded newspaper, spent compost, or coir (coconut fibre).*

Worms *Buy brandling worms or simply scoop some out of an existing compost bin. Make a hollow in the bedding and pop them in.*

Permeable membrane *An old perforated compost sack works well, or you can use a fine-gauge mesh.*

Tap *A water butt tap helps drain off the worm tea. Or drill holes in the bottom of the wormery and collect the liquid in a tray or bowl underneath.*

35cm (14in)

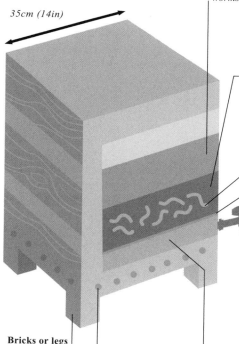

Bricks or legs *Raise the bin above ground level to increase air flow and to collect the worm tea.*

Drainage holes *Drill 7mm (¼in) holes at regular intervals around the sides of your box or bin.*

Gravel or sand *An 8cm (3in) layer of drainage material acts as a sump for surplus liquid.*

INVITING NATURE IN

Every single life in your garden, from the pollinating insects, to frogs, birds, and hedgehogs, is part of the great carbon cycle and, as such, plays a vital role in your garden's natural ecosystem.

PLANT FOR HABITAT

It may not be obvious at first why wildlife is so important to a low-carbon garden. But every single life, from the tiniest mite right through beetles, bumblebees, and butterflies to frogs, birds, foxes, and, indeed, you, is part of the great carbon cycle.

As living creatures eat (your plants and each other), excrete, and die, they continually recycle carbon-rich biomass. Ultimately, they feed it into your soil where it's locked up, then recycled again by the plants you grow. The more living things you invite into your garden, the more effective it is as a carbon sink – offsetting your carbon footprint elsewhere.

STREETS WITH **BIRD-FRIENDLY** GARDENS ARE FOUND TO HAVE ALMOST **TWICE** AS MANY **SPECIES** AS STREETS WITHOUT.

It's hard to overstate how crucial our gardens are becoming as sanctuaries for wild things, where they can escape the devastation of climate change and habitat loss. But to make the most of the benefits these creatures bring to your garden, you need them to stay. Persuade passing visitors to set up home and you'll acquire a resident population of slug-hunters, caterpillar-eaters, and aphid-guzzlers – like a built-in pest control for your garden.

SOMEWHERE TO LIVE

The low-carbon garden provides many habitats that wild creatures enjoy the most, including hedges, ponds, trees, and long grass. Choose plants carefully, too, and you can add nesting sites, egg-laying opportunities, and hidey holes for everything from solitary bees to starlings.

PLANTS FOR HABITAT:

- **Tussocky grasses:** Choose eulalia (*Miscanthus sinensis*), Mexican feather grass (*Stipa tenuissima*), or purple moor-grass (*Molinia caerulea*). Thick tussocks of ornamental grasses are great places to hide for insects; you'll also find grasshoppers, beetles, voles, and the occasional baby frog. Robins line nests with green *Miscanthus*, while brown and skipper butterflies lay eggs on purple moor-grass.
- **Evergreen shrubs:** Bay (*Laurus nobilis*), holly (*Ilex aquifolium*), and yew (*Taxus baccata*) are good choices. Sparrows love dense evergreen foliage for nesting and hiding; hedgehogs hibernate in the deep leaf litter underneath, and voles, mice, and insects make the most of the year-round shelter.

"The more living things you invite into your garden, the more effective it is as a carbon sink, offsetting your carbon footprint elsewhere."

- **Prickly plants:** Choose barberry (*Berberis*), buckthorn (*Rhamnus cathartica*), or hawthorn (*Crataegus monogyna*). Smaller birds often nest inside the safety of a prickly shrub. More than 150 species of insect live in hawthorn, and brimstone butterflies breed on buckthorn.
- **Wall climbers:** Ivy (*Hedera helix*), firethorn (*Pyracantha*), or honeysuckle (*Lonicera*) are good options. Birds nest and butterflies hibernate behind walls of dense greenery. Moths adore night-scented honeysuckle, while comma butterflies like hops (holly blues prefer ivy).
- **Hollow-stemmed plants:** Choose angelica (*Angelica archangelica*), cardoon (*Cynara cardunculus*), or hollyhocks (*Alcea rosea*). Ladybirds like dead, hollow stems – they're a safe place to hibernate out of sight of predators. Some solitary bees also nest inside hollow plant stems.

SLOW WORM SUNBATHING

Slow worms love to sunbathe. Like all reptiles, they're cold-blooded, so they have to warm up before they get going in the morning (and we all know how that feels). Provide a few hot spots and it's likely you'll attract a little colony of these handsome creatures.

Place corrugated iron sheets or pieces of roofing felt in sunny spots in the garden, among long grass so the worms won't have to cross open ground. Leave undisturbed for a few months, then carefully lift up one corner and see who's moved in.

PLANT FOR POLLINATORS

Let's be honest, gardeners have a love-hate relationship with wildlife. Birds are great when they're hoovering up aphids, but not so much when they're raiding the blackcurrants. Pollinating insects, though, are universally loved.

We all want more butterflies and bumblebees in our lives – they're charming, of course, but they also ensure our beans, courgettes, and berries produce better harvests. Unfortunately, pollinators are harder to come by nowadays. One alarming study in Germany found insect biomass – the sheer weight of the bugs around us – has plummeted by over 75 per cent in 27 years.

POLLINATING INSECTS ARE SAID TO BE **RESPONSIBLE** FOR **ONE** OUT OF EVERY **THREE MOUTHFULS** OF **FOOD** WE EAT.

Gardeners, though, can really make a difference. Fill your garden with flowers and the pollinators will come. Dandelions are at least as good as dahlias, and it's a good idea to include early-season and winter flowers to cater for year-round foragers.

SOME OF THE BEST NECTAR-RICH PLANTS

- **Apple trees:** Any spring blossom is a magnet for pollinators; they enjoy the windfall fruit, too.
- **Catmint:** (*Nepeta* x *faassenii*) It has billowing clouds of purple-blue flowers smothered in half-drunk bees all summer.
- **Comfrey:** (*Symphytum officinale*) Lush leaves are topped by bell-like flowers that are designed for long-tongued bumblebees.
- **Cardoon:** (*Cynara cardunculus*) Huge thistle-like flowers sit on 2m (6½ft) stems above silvery foliage.
- **Giant hyssop:** (*Agastache* 'Blackadder') This short-lived perennial has spires of deep blue flowers in late summer.
- **Heather:** (*Erica carnea*) A tough evergreen, it produces its pink or white bells in deepest winter for early-to-rise bumblebees.
- **English lavender:** (*Lavandula angustifolia*) This cottage-garden favourite sports fragrant blooms and evergreen silvery foliage.
- **Wallflower:** (*Erysimum* 'Bowles's Mauve') Almost never out of flower, this perennial has deep pink, nectar-rich flowers.
- **Thyme:** (*Thymus vulgaris*) Fragrant evergreen foliage erupts into tiny purple, pink, or white blooms in midsummer.

"Gardeners can really make a difference. Fill your garden with flowers and the pollinators will come."

REWILD YOUR GARDEN

A garden that is wildlife friendly can still look ornamental, with year-round colour and interest and an elegant, formal design.

Gardens are home to an astonishing diversity of wildlife. Even modest patches can accommodate a teeming throng of creatures – in one 30-year long experiment, British zoologist Jennifer Owen counted over 2,600 different species in her ordinary, suburban garden, of which almost 2,000 were insects. In fact, as long as they're managed sympathetically, without pesticides and with plenty of plants, ornamental gardens are often considerably more biodiverse than open countryside nearby. This is incredibly reassuring if you love wildlife, but don't want a wild-looking garden.

Many of the measures you can take to invite wildlife into your garden also add to its beauty and are things to aspire to in your day-to-day gardening – a long season of interest, a range of different settings, and rich planting full of flowers. If you create a garden full of variety and interest, wildlife won't take long to find it.

CREATE A WILDLIFE HAVEN

1 Holes in fence posts from 3mm (⅛in) to 10mm (½in) in diameter, will be colonized by solitary bees.

THE INVISIBLY REWILDED GARDEN

A rewilded garden encourages wild plants, insects, and other animals in, while still being beautiful.

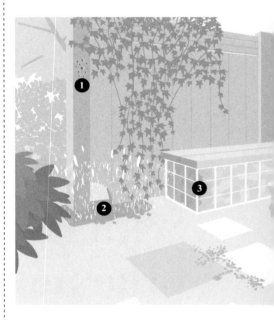

2 Holes in the bottom of fences allow a "flow" for hedgehogs and other small creatures to move in and out of the garden.

3 Wire cages filled with stones or logs (gabions) make attractive benches, and ladybirds, lacewings, and solitary bees simply love their nooks and crannies.

4 Deadwood should be left in situ – it's among the best habitat there is.

A sculptural dead tree draped in roses is a beautiful focal point.

5 Dead stems of perennials such as *Rudbeckia fulgida* var. *deamii, Veronicastrum virginicum*, and *Phlomis russeliana* look fabulous rimed with frost, so leave unpruned.

6 Dry stone walls are full of nooks and crannies for small mammals and insects, and can act as corridors between different parts of the garden.

7 Thatched roofs of wheat straw or reed roofs look beautiful on gazebos and summerhouses, plus they provide a home to all manner of insects and birds.

8 Turf roofs provide burrows for solitary bees and herbs are adored by pollinators. Add an extra layer by covering bike sheds and bin stores with greenery.

BUILD A
BUG HOTEL

Nothing says "welcome in" quite like a purpose-built, five-star hotel catering for your every need. Building a bug hotel is a fun weekend project: kids love helping, and they (and you) will also enjoy many happy hours watching new residents move in.

Many bug hotels are cheerfully scruffy, but you can also get creative, arranging materials in artistic patterns and adding other decorations until your wildlife sanctuary becomes an attractive garden feature.

MORE THAN **40 PER CENT** OF **INSECTS** WORLDWIDE ARE **DECLINING** AND **A THIRD** ARE CONSIDERED **ENDANGERED**.

The bugs won't care either way – all they're interested in is the quality nooks and crannies. The wider the variety, the more wildlife you attract: solitary bees quickly fill smaller holes, while toads and hedgehogs favour larger areas.

NATURAL HOMES

You don't have to buy anything – in fact, it's better if you don't as you can then be sure it hasn't been treated with preservatives or fungicides. Bugs don't much like plastic or glass,

which tends to attract mould and disease, or wood that's been varnished or painted. It's also best to avoid metal, which quickly gets too hot or too cold. The roof should be watertight to keep the hotel dry inside.

You don't have to accommodate a full pallet-sized hotel – mini-hotels made from recycled wood can be as small as a picture frame. And instead of one big hotel, you could have several smaller apartment blocks.

Towards the end of summer, give the whole hotel a clear-out. Replace drilled blocks of wood with new ones and refresh leaves and other organic matter. Keep your materials fresh and you'll limit parasites and diseases.

MATERIALS TO INCLUDE:

- **Bark, woodchip, and rotting wood** for beetles, centipedes, and woodlice.
- **Hollow bamboo,** dried hollyhock, or angelica stems for bee nests.
- **Wood blocks** drilled with holes in a variety of diameters from 3mm (⅛in) to 10mm (½in).
- **Clay or slate roof tiles** with larger gaps between for frogs and toads.
- **Dry leaves** and straw for ladybirds and other bugs.
- **Corrugated cardboard** rolled into tubes for lacewings and their larvae.
- **Engineering bricks** (with holes) and terracotta drainpipes for larger creatures such as frogs.

HOW TO MAKE A BUG HOTEL

Find a flat, shady location in the garden where your bug hotel guests won't be disturbed.

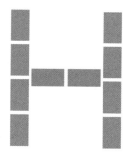

1 *Lay bricks (without mortar) on a level patch of ground in an H shape to create a base for the bug hotel to stand on.*

2 *Stack five or six untreated wooden pallets on top of each other in a pile to create a framework about 1m (3ft) high.*

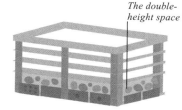

The double-height space

3 *Cut away a horizontal slat at the base on the shady side of your stack to make a double-height space, and fill to the centre with tiles or larger stones where toads and hedgehogs can hibernate.*

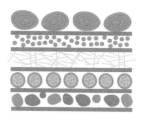

4 *Then fill in the outer gaps with a variety of different materials, aiming to create a good range of habitats to suit as many different kinds of wildlife as possible.*

5 *Finish it off with a roof of second-hand tiles, planks covered in reclaimed roofing felt, or a green roof planted with turf or wildflowers.*

POTTING COMPOST

*Mining peat for gardening releases the majority of
the carbon stores it's built up over tens of thousands
of years into the atmosphere. There are better and more
climate-friendly choices for potting compost.*

GO PEAT-FREE

If you're a gardener, especially if you grow vegetables and sow seed, the chances are you use quite a lot of potting compost. But this vast industry is destroying one of the world's most efficient natural carbon sinks – the peat bog.

Gardeners in the UK alone get through a mountainous 35 million bags of potting compost each season. European gardeners spend about €1.3 billion on it every year. That's a lot of dirt. About nine in 10 bags of compost sold contain peat. No wonder – it's a fantastic growing medium, stable, free-draining yet moisture-retentive, lightweight, and cheap.

PEATLAND AREAS WORLDWIDE SEQUESTER **370 MEGATONNES** (408 MEGATONS) OF **CARBON DIOXIDE** A YEAR.

VAST CARBON SINKS

Peat is extracted mainly from lowland raised bogs – 10,000 years' worth of carbon-rich organic matter slowly compressed underwater into spongy black earth. Peat bogs are a frontline defence against climate change, storing about 10 times as much carbon per hectare than any other ecosystem. Dig up the peat, though, and you release most of its enormous carbon payload directly into the atmosphere.

Using peat doesn't even make us better gardeners. It only really caught on as a compost ingredient in the 1970s, and we were pretty good at gardening before then. In Australia and Japan, where there are no peat bogs, gardeners have always managed perfectly well with almost no peat at all.

WEANING YOURSELF OFF PEAT

Giving up peat is one of the easier steps you can take towards low-carbon gardening. All you have to do is switch the brand you pick up at the garden centre. If the bag of compost you have in your trolley doesn't have "peat free" written clearly on the front, it isn't. "Organic" and "sustainably sourced" do not mean peat-free; nor does "eco-friendly".

Compost manufacturers are now mixing more peat alternatives into their products. But even in "reduced-peat" mixes, peat can still account for over 40 per cent of the contents. The exact proportion isn't always stated on the label, so it's best just to find a good peat-free brand and stick to it.

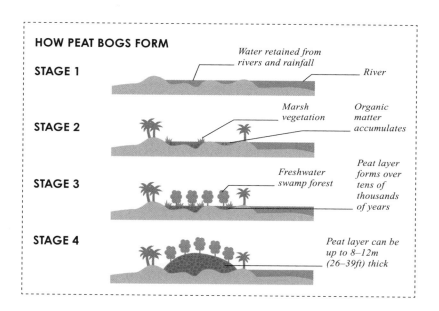

HOW PEAT BOGS FORM

STAGE 1

Water retained from rivers and rainfall

River

STAGE 2

Marsh vegetation

Organic matter accumulates

STAGE 3

Freshwater swamp forest

Peat layer forms over tens of thousands of years

STAGE 4

Peat layer can be up to 8–12m (26–39ft) thick

PEAT-FREE ALTERNATIVES

Nowadays, there are some outstanding brands of peat-free compost, even for acid-loving plants (which grow particularly well in peat). Some regularly outperform peat-based compost in trials. It's worth being fussy as there are a lot of bad peat-free composts out there too.

The best peat-free brands are so similar in performance and texture to peat-based composts they require no major adjustment of your normal growing routine. Peat-free composts based on materials easily sourced locally – such as municipal green waste, composted bark, and sheep's wool – have a lower carbon footprint than those that rely on coir, a by-product of the coconut industry that must be shipped from South Asia.

Good peat-free compost can be hard to find in many garden centres, though. If yours doesn't stock it, ask them to order it in and buy it online until they do. The power you have as a consumer is greater than you think – the choices you make in buying one bag of compost today may one day change a whole industry for the better.

BUY COMPOST
IN BULK

Even peat-free compost carries an eye-wateringly high carbon cost if you're buying it in small bags. Cut down the transport and plastic, though, and you've gone a long way towards reducing the carbon footprint of buying compost.

Compost is heavy stuff, and takes a lot of lugging about in big lorries, with the extra carbon emissions to match. Your car journeys to out-of-town garden centres bring their own contribution; and then there are all those single-use plastic sacks – made, of course, from planet-warming fossil fuels.

REDUCE THE IMPACT OF BUYING COMPOST

- **Bags for life:** Refillable "bags for life" for compost are beginning to catch on, and cut plastic use by four-fifths. The bags are still made of plastic but are heavy-duty and reusable. They hold the same volume as a sack of compost, so you can bring them back to the garden centre again and again to top up with more.
- **Bulk bags:** If you're driving to the garden centre for compost multiple times a year, getting 900-litre (198-gallon) heavy-duty plastic bulk bags delivered (each holding

the equivalent of about 18 sacks of compost) could be more environmentally friendly. If the diesel lorry is hauling your compost from a long way away, the benefits melt away. The more compost you can have delivered, and the shorter the lorry's journey, the greater the savings in carbon emissions.

- **Loose loads:** If you have the space to accommodate a loose load of potting compost, you can have your entire year's supply delivered all at once – no plastic required. As with bulk bags, source as locally as possible; you'll also need to cover your heap, or perhaps reuse your old compost sacks to bag it up until you're ready to use it.

USING OLD COMPOST SACKS

Don't throw away your old plastic compost sacks – there's plenty of life in them yet!
Growing potatoes: *Plant three seed potatoes in a 15cm (6in) layer of compost in your sack, then earth up with more compost as they grow.*
Line pots and hanging baskets: *Punch holes for drainage in the sacks, then cut to size and line containers so they don't dry out as quickly.*
Winter covers for veg beds: *Open out the sacks and spread them over veg beds, weighted with bricks, to keep soil protected from excess winter rain.*

"Cut down the transport and plastic, and you've gone a long way towards reducing the carbon footprint of buying compost."

MAKE YOUR OWN POTTING MIXES

The only truly low-carbon way to sow seed, pot on young plants, and grow in containers is not to buy compost at all, but to make your potting mixes at home.

When you make your own potting mixes, transport miles are almost zero, you use little or no plastic, and it's as satisfying as baking a good cake. Home-made blends are based on loam (garden soil), and that has a huge advantage. Plants growing in garden loam get used to its unique texture, acidity, and soil

PEATLANDS STORE NEARLY **ONE-THIRD** OF THE **WORLD'S CARBON**, AND HARVESTING THEM RELEASES IT INTO THE ATMOSPHERE.

micro-organisms – so when you plant them out in the same soil, they feel right at home. Potting mixes blend loam with more fibrous ingredients to add nutrients and improve drainage while holding on to moisture.

MIX OPTIONS

- **Compost:** Garden compost from your heap adds nutrients. It should be well rotted, without the original ingredients still visible.
- **Green waste:** Municipal green waste makes a good stand-in for garden compost; it's sterilized to kill weeds and pathogens.
- **Well-rotted manure:** Stable or farmyard manure is another high-nutrient option, but check pasture hasn't been treated with persistent weedkillers.
- **Leaf mould:** Stack fallen autumn leaves for two years in an open-sided wire bin to make organic matter fine enough for sowing seeds.
- **Composted bark:** Ask a local tree surgeon for woodchips to rot for a couple of years into soft, brown crumbs, or buy it by the bag.
- **Coir:** A good, lightweight substitute for leaf mould, but bear in mind it has a high carbon footprint.
- **Sharp sand:** Gritty sharp sand adds the drainage; buy it in unless you live near a beach (but do always ask first and rinse before using it).
- **Fertilizer:** Keep plants growing strongly by adding extra nutrients. Seaweed meal is a superb general-purpose fertilizer with a low environmental impact.

Deciding the balance of ingredients is where the magic starts. The beauty of making your own is you can have your compost exactly as you like it. Start with a basic mix, take notes, and adjust each time until you perfect it.

COMPOST RECIPES

Mix your compost as needed. You don't have to mix huge amounts at a time – use a loaded spade as the measure for a wheelbarrow load, or a trowel if all you want is a bucketful.

For seeds:
1 part loam
1 part leaf mould, bark, or coir
1 part sharp sand

For potting on and containers:
4 parts loam
1 part garden compost, municipal green waste, or well-rotted manure
1 part leaf mould
1 part sharp sand
1 heaped trowel of seaweed meal per loaded wheelbarrow
This is a basic mix. Add more sand for alpines and herbs, and use composted bracken instead of garden compost for acid-loving plants like blueberries and maples (Acer palmatum).

For cuttings:
2 parts loam
1 part leaf mould, bark, or coir
2 parts sharp sand

STERILIZE LOAM OR NOT?

You can sterilize loam by filling trays with dampened compost and heating in the oven, covered with foil, at 80°C (175°F) for 30 minutes. There are good reasons not to sterilize, though. Heating soil is carbon-heavy and will also kill off beneficial micro-organisms that could help your plants.

You'll get weed seedlings popping up in unsterilized soil, but they're easy to pull out, and disease is rarely a problem if the compost is freshly made.

LOW-CARBON CONTAINER GROWING

If you live in an urban environment, container planting is often the only outside option for small gardens and balconies, alongside houseplants. But plants in pots are reliant on you for everything – so your growing choices do matter.

Although potted plants have a smaller role to play than plants in the ground, they still take carbon from the air. Indoors, they'll also help create a more natural environment by absorbing pollutants and oxygenating the air. Manage them sustainably and you'll maximize those benefits.

SUSTAINABLE OPTIONS

- **Use second-hand containers:** Look in reclamation yards, vintage shops, and online auctions for zero-carbon pre-loved planters. Or build your own planter boxes from reclaimed wood.
- **Choose the biggest you can:** With a few exceptions, the larger your pot,

the more your plants can spread out their roots. Choose pots that have a diameter of at least 45cm (18in).
- **Grow plants that like container life:** Plants that are happiest in containers are relatively slow growing, with naturally small root systems, such as dwarf ("patio") fruit and veg, small shrubs like skimmias, and herbs.
- **Use pot dishes:** Catch and recycle water and nutrients by using containers with saucers or reservoirs wherever possible.
- **Plant permanently:** Permanent shrubs, trees, perennials, and grasses absorb carbon over the longest period and generally need less compost, water, and fertilizers than hungry, fast-growing annuals.

IF YOU WANT ANNUALS...

- **Grow from seed:** Raise your own annual flowers, bedding plants, or vegetables from seed, sown into pots of home-made compost.
- **Don't change all the compost:** When replanting annuals in spring,

ANNUAL **CARBON DIOXIDE** EMISSIONS FROM **PLASTIC** COULD GROW TO MORE THAN **2.75 BILLION TONNES** (3 BILLION TONS) BY **2050**.

> *"Although potted plants have a smaller role to play than plants in the ground, they still take carbon from the air."*

leave as much compost undisturbed as possible by digging out and replacing only the top 5cm (2in) of old compost.

- **Use captured rainwater:** Plants prefer rainwater, as it's full of nitrogen and slightly acidic (unlike tapwater, which tends to be limey). You'll avoid all the emissions involved in treating and transporting tapwater, too.
- **Feed with home-made fertilizers:** Artificial fertilizers are another source of greenhouse gas emissions (and tricky-to-recycle plastic bottles) so make your own (see p.106).
- **Grow a container fruit garden:** Potted orchards of fruit trees and berries are the ultimate in sustainable container growing, majoring on permanently planted, carbon-rich woody trees.

ORCHARD FRUITS

The following rank among the best container orchard fruits:
- Apple Ballerina®
- Peach 'Bonanza'
- Cherry on dwarfing Gisela 5 rootstocks
- Quince and Sibley's Patio Medlar
- Figs
- Blueberries (planted into acidic ericaceous compost)
- Chilean guava (*Ugni molinae*)
- Raspberry 'Ruby Beauty'
- Blackberry 'Dart's Black Cascade'
- Strawberries

CARING FOR YOUR POTTED ORCHARD

- **Buy bare root plants** if possible (see p.158), and pot up in peat-free, ideally home-made, compost while still dormant.
- **Keep well watered** with rainwater.
- **Add a general-purpose feed** weekly in summer; switch to high-potassium comfrey tea (see p.106) once flowers form.
- **Replenish the top** 5cm (2in) of compost each year. Every other year, tip the whole plant out and cut away about 30 per cent of the roots before returning to the pot with fresh compost.

SEED SOWING

Sowing a seed and watching it burst into life never loses its thrill if you're a gardener. But sowing can carry a surprisingly high carbon footprint – unless you do it the low-carbon way.

MAKE YOUR OWN POTS, TRAYS, AND MODULES

Seed sowing is one of the most exciting times of the gardening year, but it's all too easy to bring home mountains of single-use plastic from the garden centre. Reuse old plastic plant pots, trays, and modules where you can, or just avoid it altogether.

Spring has sprung, the sun has some real warmth, and at last you can get the new season under way! At this thrilling time of year, you get to experiment with new varieties, revisit old favourites, and start off whole new planting schemes. And that excitement you get when the first tiny seedling leaves poke above the compost never, ever fades.

DIRECT SOWING

Sowing home-saved seeds straight into shallow drills in the garden is the zero-carbon option. It generally gives the best results, too, as you're raising plants exactly where they will grow, without disturbing the roots.

But emerging seedlings have to dodge an army of hungry slugs and mice, and before mid-spring the soil has rarely hit the 10°C (50°F) minimum required for germination. You can cheat by prewarming the soil – place a glass barn cloche over the top two weeks before sowing. But your growing season is still likely to be shorter, and more beset by pests, than if you sow into pots and trays under cover.

Seed sowing in artificial conditions is highly intensive gardening, though. If you use new plastic trays, pots, and modules (often low grade and short lived for seed sowing), commercial peat-based seed compost, and a propagator heated to a cosy 18–21°C (64–70°F) so you can raise tomatoes in late winter – you quickly rack up an eyewatering carbon footprint.

Luckily, you can still sow seeds under cover, even early in the year, without resorting to any of this.

APPROXIMATELY **500 MILLION PLANT POTS** AND **SEED TRAYS** ARE SOLD IN THE UK EACH YEAR. THE MAJORITY ARE **INCINERATED** OR SENT TO **LANDFILL.**

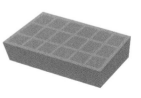

TAKING THE PLASTIC OUT OF SOWING SEEDS

There are good reasons not to sow into plastic, and they aren't all about the carbon cost; though, for new plastic, made from fossil fuels, that's sky high. Most plastic plant pots these days are made from about 80 per cent recycled plastic, but as plastic can only be recycled once or twice, it ends up in landfill, incinerators, or the oceans eventually.

"All plastic – even recycled – ends up in landfill, incinerators, or the ocean eventually."

PLASTIC POTS TAKE ABOUT **450 YEARS** TO **DECOMPOSE**.

But there are good alternatives to plastic and they are better for your plants. The roots of seedlings grown in porous wood, paper, and cardboard, for instance, grow better. The roots also grow straight through the sides instead of circling into ever-tightening knots. So, you get healthier plants that establish faster and don't contribute to global warming. Choose from among these low-carbon options for seed sowing.

- **Wooden seed trays:** These are no more expensive than plastic if you buy second-hand from auction sites. Or make your own from scrap wood – those are free.
- **Newspaper pots:** You can buy a wooden gadget to make them, or use straight-sided shot glasses to roll strips of newspaper into handy 5cm- (2in)-diameter modules, ideal for sowing and pricking out.
- **Toilet roll tubes:** Save the cardboard tubes from your loo rolls, stand them in a seed tray, and fill with compost to use for sowing large seeds such as beans.
- **Soil blocks:** Compressed blocks of home-made compost make ideal zero-waste seed-starter modules to plant out directly.
- **Cardboard pots:** For larger 10cm (4in) pots, make a box from thin cardboard, held together with paper masking tape.

MAKE LEAF MOULD

What you put in your low-carbon, biodegradable pot is as important a consideration as the pot itself. Leaf mould is easy to make, can be used at any time of the year, and is completely free.

Unless you live near a particularly enlightened garden centre, every bag of seed compost you'll find on sale contains peat. Some good peat-free brands offer seed compost, by mail order only. It's always worth asking your garden centre to stock peat-free seed compost, of course – if they

---

94 PER CENT OF THE UK'S LOWLAND RAISED **PEAT BOGS** – ONE OF ITS RAREST HABITATS AND MOST IMPORTANT CARBON STORES – **HAVE BEEN LOST.**

---

know you'll buy it, they'll find it! But in the meantime you can make your own for absolutely no cost at all.

LOW IN NUTRIENTS

Leaf mould is one of gardening's best-kept secrets. Even gardeners don't always know about it, and cart off their autumn leaves to the tip, or (worse) burn them without realizing they're sending one of the garden's most valuable resources up in smoke.

Autumn leaves are carbon-rich lignin (a wood component), left over after the nutrients from the living leaf have been absorbed back into the tree as it goes into hibernation in autumn. Once they have rotted, you're left with organic matter that's stable, low in nutrients, and fibrous (so free-draining), yet absorbs water like a sponge, increasing the ability of your soil to hold on to moisture by almost 50 per cent. All of that makes it an ideal medium for germinating seeds.

WHICH LEAVES?

You can use any kind of leaf to make leaf mould, but thinner leaves, such as ash and beech, break down faster than thicker ones, such as sycamore and horse chestnut. If you have a lot of thicker leaves, shred them roughly before adding them to the bin.

Pine needles also work well but they break down more slowly (allow three years). Other types of conifer and evergreen leaves from holly and rhododendrons take so long to break down they're more useful shredded for general compost.

Rough leaf mould, suitable for mulching and soil conditioning, is ready after about a year. But for fine-grade leaf mould, suitable for sowing seeds, you'll need to wait at least another year for it to break down into black, soft fine crumbs.

HOW TO BUILD A LEAF-MOULD BIN

Leaves rely mainly on fungi to decompose, unlike home-made compost in which bacteria do all the hard work. This means leaf mould requires quite different conditions, so it's best to rot it down separately, in its own bin.

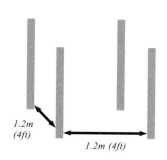

1.2m (4ft)

1.2m (4ft)

1 *Drive four 1.5m (5ft) wooden posts into the ground in a square. Leaf mould bins can be as big as you like, but no smaller than 60cm (2ft) square; a 1.2m (4ft)-square size works well.*

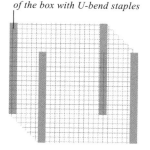

Fix the wire mesh to each side of the box with U-bend staples

2 *Staple 25mm (1in) gauge wire mesh (ideally reclaimed or reused) firmly to the posts to make an open-topped box.*

Water to keep damp, if needed

3 *Collect leaves from late autumn and pile them in the bin. Tread them down firmly as you go, and keep them damp.*

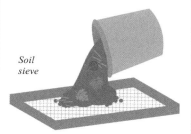

Soil sieve

4 *Once the first bin is fully rotted, detach the wire and empty out the dark, crumbly contents. Pass through a soil sieve and use neat for sowing seeds. Refill your leaf mould bins as they empty for a constant supply.*

RAISE SEEDS INDOORS

Some plants, including tomatoes, peppers (both sweet and chilli), and aubergines, need a long season to reach maturity and produce fruits, so it's essential to sow them as early in the year as possible. If you live in a cooler part of the world, that means artificial heating in an indoor environment.

Heating a traditional greenhouse solely to raise seedlings is prohibitively expensive – in terms of carbon emissions and your electricity bill. A thermostatically controlled heated propagator targets warmth much more precisely and is more economical, but if it's in your greenhouse, it must work hard to raise the ambient temperature on very cold days, bumping up its energy use.

Better than either of those options is to use the ample heating you're already spending energy, carbon emissions, and money on, and sow seeds and raise seedlings indoors.

MOST **SEEDS GERMINATE** WHEN THE **SOIL TEMPERATURE** IS BETWEEN **12** AND **21**°C (**54** AND **70**°F).

Room temperature averages 18–21°C (64–70°F), which is ideal for germinating most seeds, so they shouldn't need extra heating at all. The only thing seedlings growing inside may need help with is light.

WINDOWSILLS AND GROW LIGHTS

You can buy off-the-shelf grow light kits for raising seeds indoors, or just make your own system by suspending LED grow lights over trays of seedlings on a shelf.

A bright south- or south-west-facing windowsill is a popular place for germinating seeds inside, but it's not ideal as the light source comes from one direction, encouraging seedlings to grow leggy and weak. Turning them by a quarter turn each day helps; you can also put them in a light-reflecting box to bounce the light around a little more.

Find a cardboard box that fits on your windowsill, then cut out the side that will face the window. Line the whole thing with foil and put your seedlings inside. Place on the windowsill (with the back towards you) and the foil reflects the light from all sides, creating a much more naturally lit environment.

If you don't have windowsills, you can still raise your seedlings indoors using low-energy LED grow lights.

SETTING UP A SEEDLING TRAY
WITH GROW LIGHTS

Place seedling trays on a sturdy frame and suspend LED grow lights over the top to create the ideal growing conditions for seedlings indoors.

LED grow light *kits are readily available online and easy to set up. Hang them above each shelf, about 30cm (12in) from the top of your seedlings.*

Use a sturdy frame *to hold trays and grow lights. Any shelving unit will do, or you can make your own out of wood. Place it near an electricity socket so you can plug in the lights easily.*

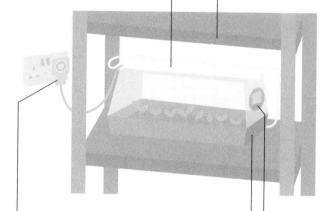

Plug the lights *into a socket via a timer to regulate the amount of daylight you give your seedlings. Programme the lights to come on for 10–12 hours during the day, then switch off at night to give the seedlings some darkness – plants need to rest too!*

Use a waterproof tray *to hold the seedling pots. If the room regularly drops below 10–15°C (50–59°F) at night, you can add a heat mat or soil-warming cables in sand for a temporary boost in temperature.*

A max–min thermometer *gives you the highest and lowest temperatures since the last reset, so you can monitor what's happening and add extra heat if it's required.*

MANAGING SOIL

One of the best ways to protect your garden soil's carbon-sink potential is to disturb it as little as possible. That means stop digging – and ditch the artificial fertilizers.

THE WORLD
BENEATH YOUR FEET

You might think that most of the carbon in your garden is stored in the plants, but plant biomass accounts for less than a fifth of your garden's carbon sink potential. The rest is under your feet.

The soil you grow your plants in holds 83 per cent of your garden's carbon, sequestered in plant roots, fungi, bacteria, living and dead animals, and humus (decomposed plant matter). Gardens are particularly good at storing carbon as they're richer in plants, and therefore organic matter.

So, as gardeners we are already looking after a significant carbon sink. Care for your soil, and support the teeming ecosystem it contains, and you can boost your garden's carbon stores even more.

BOOSTING YOUR BACKYARD CARBON SINK

- **Stop digging:** Turning soil exposes carbon to the air.
- **Avoid treading on your soil:** Walking on soil compacts it and drives out the air.
- **Plant permanently:** Major on trees, shrubs, and long-lived perennials.
- **Grow lots of plants:** Keep paving and buildings to a minimum.
- **Keep the soil covered:** Bare topsoil is eroded by exposure to weather, taking its load of carbon with it.

THE SOIL FOOD WEB

Leaving your garden's soil undisturbed also keeps its delicate ecosystem intact.

***Nematodes** are microscopic worm-like creatures that feed roots via their nutrient-rich faeces.*

***Mycorrhizal fungi** plug plant roots into the soil's nutrient bank; there's evidence they help "transmit" signals between plants.*

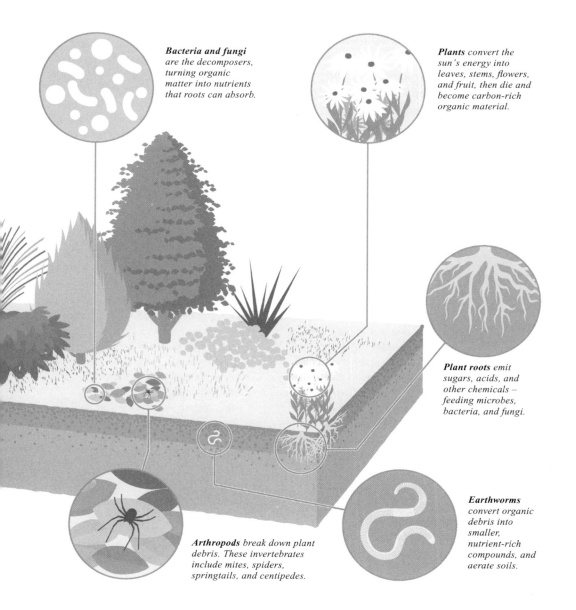

Bacteria and fungi are the decomposers, turning organic matter into nutrients that roots can absorb.

Plants convert the sun's energy into leaves, stems, flowers, and fruit, then die and become carbon-rich organic material.

Plant roots emit sugars, acids, and other chemicals — feeding microbes, bacteria, and fungi.

Earthworms convert organic debris into smaller, nutrient-rich compounds, and aerate soils.

Arthropods break down plant debris. These invertebrates include mites, spiders, springtails, and centipedes.

COVER
THE GROUND

While you're building up the life in your soil, it's also important to make sure the carbon it's already storing stays put. Covering your soil not only keeps carbon locked up, but also prevents erosion and suppresses weeds.

Soil left bare and open to the elements is vulnerable, especially in the world of dramatic weather extremes brought on by climate change. A torrential downpour can simply wash soil away, along with the essential nutrients it contains for your plants; extreme heat dries it out before winds whip it into dust storms. With your topsoil goes all its payload of sequestered and stored carbon, as well as the goodness it holds.

Intensive agriculture and annual ploughing have reduced the depth of topsoil in parts of the US by more than 50 per cent over the last century. In 2015 the United Nations (UN) estimated the world's topsoil is so depleted it can support just 60 more

---

ACCORDING TO THE **UN**, THE WORLD'S **TOPSOIL** IS NOW SO **DEPLETED** IT CAN SUPPORT JUST **60** MORE **HARVESTS**.

---

harvests. Gardens have a crucial role to play in providing pockets of deep, rich, carbon-storing soil.

Nature's way of protecting soil, as we gardeners know only too well, is to send in the weeds. So, any time you see a weedy patch, it's a warning sign your soil is less well protected than it should be, and it's time to cover up.

GROUND-COVERING TACTICS

- **Mulches:** Spread a 5cm (2in) mulch of organic matter over your soil and keep it topped up to feed the soil and hold in moisture, as well as protecting it. Mulch boosts soil's carbon content by keeping the underground ecosystem healthy.
- **Dense planting:** In a garden packed with plants, there's no room for weeds. Fill spaces between new plantings with direct-sown annuals; in the veg patch, keep a steady supply of young vegetable plants ready to drop into gaps as they open up.
- **Living mulches:** Low-growing, spreading plants act as "living mulches", protecting the soil, holding in moisture, and keeping out the weeds. A carpet of ground-cover plants does the job under ornamental plants; in a vegetable garden, plant squash or strawberries between taller crops.

"Gardens have a crucial role to play in providing pockets of deep, rich, carbon-storing soil."

USE GREEN MANURES

When you have a large area of soil to keep covered and planting isn't practical, there is one way you can both cover your soil and feed it with nutrient-rich, carbon-packed organic matter.

Green manures, or cover crops, are very fast-growing, high-biomass plants grown specifically to improve the soil. Simply scatter their seeds over the area you want to cover, rake them in, and keep watered. Within weeks, your soil will be blanketed in lushly protective greenery.

GREEN MANURES ARE COVER CROPS INCLUDING PLANTS FROM THE **LEGUME**, **MUSTARD**, AND **GRASS** FAMILIES.

Summer manures can be managed with less soil disturbance (and therefore carbon release). Simply let them die back naturally and leave the dead top growth in place over winter. Rake off before sowing and add to the compost to return to the soil as a mulch. Winter cover crops, however, require deep cultivation as you turn the roots into the soil a few months before sowing.

Any green manure will both protect your topsoil and pump it full of nutrients, which in itself boosts not just your crops but also your soil's ecosystem.

SUMMER GREEN MANURES

Sow these summer green manures from spring to midsummer on after clearing early crops:

- **Buckwheat** (*Fagopyrum esculentum*) grows lush and leafy with succulent foliage and masses of white flowers, loved by pollinators.
- **Crimson clover** (*Trifolium incarnatum*) is a natural fertilizer as it fixes nitrogen from the air as it grows.
- **Fiddleneck** (*Phacelia tanacetifolia*) has masses of powdery blue, nectar-rich flowers; it self-seeds like mad if you let it, though.

WINTER GREEN MANURES

Sow these winter green manures from late summer to early autumn:

- **Winter-hardy salads** such as *Claytonia*, chicory, and corn salad grow quickly into mats of nitrogen-rich rosettes; in spring, shear off the top growth and add to the compost bin then dig in the roots to rot down.
- **Common vetch** (*Vicia sativa*) can be sown well into autumn and injects nitrogen into soil as it grows.
- **Field beans** (*Vicia faba*) are closely related to broad beans and also fix nitrogen from the air; they're edible, so you'll get an extra harvest.

HOW GREEN MANURES WORK

Choose your green manure based on how it can improve your soil and what a particular patch needs.

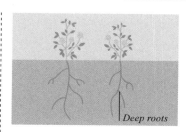

Nutrient sharing *Deep-rooted plants, such as alfalfa and fodder radish, can draw up nutrients from lower levels of soil and make them available to plants at the surface.*

Lush growth *All green manures produce lots of green biomass in a short time. Add them back to the soil, via composting or rotting in situ, to raise the soil's nitrogen levels, too.*

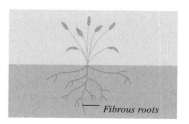

Aeration of soil *Manures with fibrous roots, including field tares and grazing rye, break up the soil, helping to lighten claggy clay.*

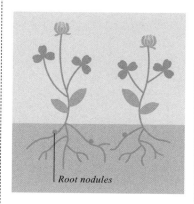

Nitrogen-fixing nodules
Leguminous green manures, such as clover and field beans, use bacteria in their root nodules to "fix" nitrogen from the air into the soil.

Pollinator friendly *Flowering green manures, such as Phacelia and buckwheat, attract pollinating insects, boosting both your vegetable harvest and your garden's biodiversity.*

PEST CONTROL

Pests are a major problem, but synthetic pesticides and chemicals can be harmful to the environment and destructive to biodiversity. There are plenty of alternatives for the low-carbon gardener to try.

PREVENTION IS BETTER THAN CURE

Lots of creatures, large and small – and very tiny indeed – enjoy eating your home-grown veg as much as you do. They're partial to ornamental plants, too. But resist the temptation to reach for the pesticides.

Resorting to pesticides to zap an army of pests might seem easy and effective, but pesticides are second only to fertilizers in their ability to inflict collateral damage in the form of pollution and climate change. They are also one of the main suspects for the widespread and dramatic loss of insect biodiversity in recent years.

It's not just the carbon footprint involved in making them, though that's high enough – pesticides account for about 9 per cent of energy use in arable farming globally.

Many insecticides kill bugs indiscriminately, including those that would have eaten your pests for you, had you let them. They leave your

PESTICIDES ARE FOUND IN **90 PER CENT** OF **STREAMS AND RIVERS** IN THE US.

plants with no natural predators left to defend them, and your garden devoid of the insect life and, therefore, carbon-rich biomass it needs to thrive.

Neonicotinoid pesticides such as acetamiprid (largely banned in the EU, though not yet in the US) are absorbed entirely into a plant, even appearing in low amounts in its pollen. Acute exposure has been proven to impair bees' ability to forage – perhaps one explanation for the average 40 per cent annual decline in US honeybee population.

Organic pesticides are much less toxic. That said, some, including pyrethrins and insecticidal soaps, also don't discriminate between pests and pollinators, so only use them if you really have to. Minimize damage by following the manufacturer's instructions to the letter and spraying late in the evening when pollinators are less likely to be active.

In any case, by the time you see bugs on your plants, they're already doing damage. Resist the spray gun and instead put some defences in place before there's even a problem.

"Pesticides inflict pollution, climate change, and the destruction of biodiversity."

Pests come in all shapes and sizes.

NATURAL SOLUTIONS

- **Patrol regularly:** Take your cup of tea outdoors each day and walk around your plants – you'll quickly spot the tiniest change and catch infestations in their early stages.
- **Use barriers:** Surround plants with a physical barrier and pests don't get a look-in. Mesh cages keep out flying insects, including cabbage-white butterflies, pigeons, and aphids. A plant perimeter of heaped-up bran stops slugs, as well.
- **Keep plants growing strongly:** Just as you're more prone to catching colds when you're run down, plants are affected more by pests and diseases when they're not growing well. Choose your plants carefully, plant them in the right place, and they'll stay in tip-top condition and largely pest free.
- **Encourage natural predators:** Welcome in wildlife and your garden becomes a naturally balanced ecosystem with built-in pest control. Aphids on your roses become food for blue tits and ladybirds, and frogs hoover up slug eggs and young slugs by the dozen.
- **Keep clean and tidy:** Viral and bacterial diseases and fungal infections are easily spread, carried on secateurs, lurking on yellowing leaves, or among leftover compost in unwashed pots. So, keep things neat and tidy and disinfect pruning blades between plants.
- **Grow resistant varieties:** There are dozens of new vegetable varieties with resistance to disease, such as blight-tolerant 'Sarpo Axona' potatoes. Near-wild "species" plants, such as *Rosa gallica* roses, also tend to be more robust.

MIX IT UP

If you grow your plants the traditional way, in neat rows of the same species, you may as well hang a neon sign overhead saying "Come and get me!" Mixing two or more plants will attract a wider range of insects – including natural predators to help keep pests under control.

Pests can sniff out a likely target from some distance. Organize plants into an appetizing row, helpfully cleared of competing greenery, and the scent brings in every bug for miles around.

Jumble up different kinds of plants, though, and pests find it dreadfully confusing. They can't distinguish their target scent as easily among lots of highly aromatic herbs. It's harder to find the right foliage, too, when it's hiding among distractingly bright yellow *Anthemis* flowers. You also bring in more beneficial insects – a lacewing might be drawn in by a pot

- -

PLANTING **POLLINATOR-FRIENDLY BORAGE** ALONGSIDE **STRAWBERRIES INCREASES YIELDS** BY OVER A THIRD.

- -

marigold, but it'll stay to eat the aphids on a neighbouring lettuce.

Polycultures, in which vegetables rub shoulders with flowers, fruit, and herbs, spread the risk: slugs may eat one or two lettuces, but they can't obliterate an entire row.

There are many other benefits to companion planting. The soil stays covered, and nutrients are shared by a variety of plants with different needs. You can double the return from a single vegetable bed by interplanting taller crops, such as Brussels sprouts, with shade-loving lettuces, parsley, or chard. Wide-leaved squashes help to hold moisture in the soil and keep out weeds; and nitrogen-fixing legumes, such as broad beans, feed neighbouring spinach.

COMBINATIONS TO TRY

- **Sweetcorn, squash, and French beans:** Plant sweetcorn first, then, once it's growing stronger, sow beans to fix nitrogen and scramble up the corn. Squash acts as a living mulch, keeping soil damp and weed free.
- **Cabbages and red clover:** Growing clover under cabbages results in a 60–90 per cent reduction in egg-laying cabbage butterflies and moths. Nitrogen-fixing clover feeds the soil, too.
- **Strawberries, borage, chives, and thyme:** Strawberries cover the ground, with a bumper crop as pollinators are attracted by borage and thyme flowers. Pungent chives help repel aphids.

"Jumble up different kinds of plants, and pests find it dreadfully confusing."

MAKE YOUR OWN PESTICIDES

Sometimes, despite your best efforts, pests break through your defences. By the time you've noticed, they are sizeable colonies. At this stage you have little choice but to fight back. Harness natural predators and use home-made repellents and you can still win the war on pests.

Pests are astonishingly successful at reproducing, and fast. Aphids, for example, can more than triple their populations every day. It's not surprising it's hard to keep up.

BIOLOGICAL CONTROLS

One solution is to use a biological control. These natural predators are supplied live or in a state of suspended animation and you release them in your garden. They quickly multiply to match the pest population and can be very effective in bringing the situation back under control.

Most biological controls are for greenhouse use, but some can be used outside – *Phasmarhabditis hermaphrodita*, a nematode that preys on slugs, is simply watered on to the ground in spring. They are expensive, though, and biocontrol production is an international business with a carbon footprint to match. You can take the carbon out of the equation altogether, though, by making your own pesticides at home.

HOME-MADE PESTICIDES

You can use materials you probably already have at home:

- **Tomato leaf deterrent:** Chop 500g (1lb) of tomato leaves and steep overnight in 1 litre (1¾ pints) of water. Strain and spray. Tomato leaves contain toxic alkaloids that deter, and sometimes kill, sap-sucking insects.
- **Beer slug traps:** Pour 5cm (2in) of beer into a large yoghurt pot and bury it close to vulnerable crops, leaving a lip above the ground so beetles can't fall in. Slugs adore beer but once they crawl in, they can't get out again.
- **Soapy water:** Stir one tablespoon of liquid Castile soap into 1 litre (1¾ pints) of warm water. Spray onto leaves to coat and suffocate sap-sucking insects.
- **Milk:** Dilute milk (full fat or semi-skimmed) 40:60 with water and spray on courgette, squash, and grapevine leaves to treat mildew. It's thought the lactose reacts with sunlight; in some trials, this outperformed artificial fungicides.
- **Rhubarb water:** Boil 500g (1lb) of rhubarb leaves in 1 litre (1¾ pints) of water for half an hour, strain, and spray. High in toxic oxalic acid, it also kills beneficial insects, so use with care, only on ornamental plants as it's toxic to humans, too.

HOW TO MAKE PEPPER
AND GARLIC SPRAY

Crushed garlic releases unpleasant sulphur compounds in self-defence; chilli peppers contain fiery hot capsaicin, too. Make either or both into a spray and you'll render your plants unpalatable to even the most determined of pests.

Use bulbs whole

The muslin filters out the bits

3 *Strain the liquid through a muslin into a clean bowl, then add a tablespoon of liquid Castile soap. Decant into a clearly labelled bottle.*

1 *Take two whole bulbs of garlic or a generous handful of fresh, hot chillies and blitz briefly in a blender.*

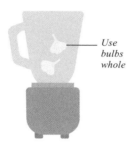

2 *Tip the chopped vegetables into a 1-litre (1¾-pint) preserving jar and fill with water. Steep the mixture overnight, or for up to a week for chilli spray.*

4 *Pepper spray can be used neat, but dilute garlic sprays 1:4 with water. Choose a dry day and spray both sides of the leaves thoroughly. Repeat weekly.*

BUYING PLANTS

*The plants bought in a garden centre often carry high
hidden carbon emissions. Do your research before
purchasing and choose plants with the environment in
mind. It could be one of the most significant contributions
you can make as a climate-friendly gardener.*

BUY LOCALLY

Your choices when buying new plants are as significant for your environmental impact as almost anything else you do. So, next time you're tempted to load up your trolley at the garden centre, ask yourself what those plants really cost, as it won't be on the tag.

We gardeners love a bit of retail therapy. Whether you're a regular at your local garden centre, haunt specialist nurseries, or binge buy at flower shows, chances are you, too, will have felt the rush of happiness new plants can bring. The average British gardener spends about £10,000 on plants over a lifetime. American households fork out more than $500 a year on their gardens, while in the EU, where almost half the world's pot plants are produced, the market is worth over €22 billion a year.

Plant production can be surprisingly international. Your impulse buy might have started life as breeding stock in Germany and been shipped to Ethiopia to grow on. Cuttings might be air-freighted to the Netherlands for rooting, then back to Germany for

potting up, before travelling by ship and lorry to your local garden centre.

The UK gardening industry is working to improve its environmental impact. The use of biological and cultural controls to tackle pests is now more widespread, and you're more likely to find plants grown in peat-free compost on sale – almost a quarter (23 per cent) of compost used by growers in 2019 was peat-free (if your garden centre doesn't sell peat-free plants, ask why not). But there's still a long way to go. The majority of plants sold are still raised in peat-based compost, passing through several sizes of plastic pot on the way.

SOURCING PLANTS SUSTAINABLY

It's OK – you don't have to give up buying new plants! Lowering your carbon footprint when stocking your garden borders simply requires you to ask a few searching questions.

- **Where does it really come from?** "Home grown" or "grown in a British nursery" doesn't always mean the plant originated in Britain. Technically, it just means the plant

THE AVERAGE BRITISH GARDENER SPENDS ABOUT **£10,000** ON **PLANTS** OVER A **LIFETIME**.

has spent "considerable time" in a UK nursery; it could have come from cuttings shipped from Uganda.

- **Did you grow it yourself?** Smaller, specialist nurseries often propagate their own stock on their own premises. Some garden centres, too, work in partnership with local growers to raise plants with as few plant miles as possible.

- **Has it been sprayed with pesticides?** You're unlikely to find pesticide-free plants in a garden centre, where commercial considerations dictate low-cost, catch-all solutions for keeping stock free of pests. But there's now a number of smaller nurseries raising plants organically, therefore without artificial pesticides. Search for "pesticide-free plant suppliers" on the RHS website.

- **Is it grown in peat-free compost?** Commercial UK growers are under notice from the government to stop using peat by 2030 to help meet climate emissions targets. But you're still unlikely to find plants raised in peat-free compost at mainstream garden centres. Again, smaller nurseries are leading the way, along with organizations such as the National Trust.

BUYING MAIL ORDER

Finding plants raised locally and organically, in peat-free compost, can be a tall order. Nowadays, though, we have the power of the internet at our fingertips.

A plant couriered to your door from a nursery three counties away that grows its own organic, peat-free stock can often have a lower carbon footprint than the one you drove to your nearest garden centre to buy.

Track down suppliers who wrap plants in paper, pad boxes with shredded cardboard, and use compostable bioplastic bags instead of plastic pots.

BUY WITH
THE SEASONS

Skip the plastic containers and peat-based potting compost and choose bare-root plants (dug up while dormant in late autumn), simply wrapped in newspaper or hessian or packed into boxes. Plant immediately and they'll have all winter to build a good root system before bursting into life in spring.

Once, buying plants was as seasonal a garden activity as picking strawberries. Gardeners a century ago would tour flower shows in summer just like we do today, but they were there not to buy plants but to make lists. Nurseries would show off their choicest new stock, hoping to catch the eye of head gardeners – if they liked what they saw, they noted it down and would often preorder.

BARE-ROOT PLANTS ARE **LESS** EXPENSIVE TO **TRANSPORT** THAN **ESTABLISHED** PLANTS, **LOWERING** THEIR **CARBON FOOTPRINT.**

Then in autumn, everyone would go on a plant-buying frenzy. Reams of plant names would arrive at each nursery, each order carefully dug up from the fields bare root and dormant, packed into boxes, then despatched for planting straight away.

BACK TO A BYGONE ERA

Nobody dreamt of buying in midsummer – it's quite the worst time to plant as the soil is at its driest and plants are in full leaf, putting real stress on roots limited by growing in pots. Instead, you made the most of the golden months of autumn, when the rain softens warm soil and plants can settle in more quickly.

The old ways have largely fallen out of fashion in the age of the cheap, mass-produced plastic pot. But they haven't disappeared altogether. And now bare roots are back in demand as concerns rise over plastic use and peat-based container production.

You can buy almost any type of plant bare root – from raspberry canes and roses to bundles of wallflowers, plus fruit trees, hedging, and some trees and shrubs. An ever-increasing range of perennials are available bare root, too, including peonies, irises, hardy geraniums, and verbascums.

WHEN THEY ARRIVE

The bare-root season runs from autumn to spring. Unwrap bare-root plants as soon as they arrive and soak in water for an hour before planting straight away. If you can't plant immediately, bury them temporarily in loose soil or compost. It's about as sustainable as plant buying gets – and you'll look forward to it all year.

"Bare-root plants are about as sustainable as plant buying gets – and you'll look forward to it all year."

RAISE YOUR OWN PLANTS

The only way to be certain of getting new plants with the lowest possible carbon footprint for your garden is to become your own plant nursery. Producing plants from home-saved seed, cuttings, and divisions gives you complete control over where they come from and how they're raised.

You get a whole new border full of beautiful plants for nothing, in the happy knowledge that it has barely any cost to the environment either. It takes patience, but the sense of achievement when you know the flourishing plant in your garden is all down to you is hard to beat.

Increasing stocks of plants you already have is straightforward, but to expand your range, you have to be inventive. Friends and relatives may be happy for you to take cuttings or seeds from plants in their gardens, perhaps in return for one of your own plants. Membership seed schemes, like that offered by the RHS, are also happy hunting grounds for more unusual varieties. They offer a selection of seeds each year, collected from their own or members' gardens.

MAKING DIVISIONS

Divisions are the easiest way to propagate perennials and grasses, and they often flower in the first year after planting. Any mature clump-forming perennial will divide into smaller plants.

Dig up the clump in autumn or spring, and gently prise it apart into fist-sized chunks. You may need to cut up bigger clumps with a sharp spade. Replant each division where you want it to grow.

RAISING FROM SEED

A handful of seed gives you more plants than you could ever afford to buy. Many perennials are easy from seed, including purple top (*Verbena bonariensis*), purple coneflower (*Echinacea purpurea*), and Balkan clary (*Salvia nemorosa*).

Sow into trays of home-made seed compost (see p.134) in spring under cover, at 18–21°C (64–70°F). Prick out and pot on seedlings as they grow, and plant outside from early summer.

TAKING CUTTINGS

Collect softwood cuttings from new spring growth, semi-ripe cuttings in summer, or hardwood cuttings (from shrubs) in winter. Most take two or three years to reach maturity.

Cuttings must have good drainage, so fill pots with a 50:50 mix of home-made multipurpose compost and sharp sand. Hardwood cuttings go straight into the ground – bury to two-thirds their length in gritty soil.

HOW TO TAKE
SEMI-RIPE CUTTINGS

Semi-ripe cuttings are really quick to root as they're made in midsummer when conditions are just right. They are suitable for a wide range of plants, including evergreens, perennials, and climbers.

1 *From late summer, look for strong, non-flowering sideshoots, firm at the base but with a soft tip. Cut away a 10–15cm (4–6in) length.*

3 *Make six holes in a 10cm (4in) diameter pot of gritty compost with a pencil. Drop each cutting in to about half its length. Firm in gently and water.*

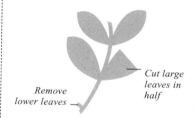

Cut large
leaves in
half

Remove
lower leaves

2 *Trim away the stem to just below the lowest pair of leaves, remove the growing tip, and strip away lower leaves so there's two to three pairs of leaves left. Trim larger leaves by half.*

4 *Pop your cuttings in their pots into a propagator; this warmth encourages rooting. You should have new young plants within six to eight weeks.*

SHOPPING FOR THE GARDEN

Shopping for second-hand gardening gear, buying locally, or even finding what you need within your own garden walls, are just some of the ways to guarantee that your garden supplies are as low carbon as possible.

EMBRACE
THE OLD

Wander into your local garden centre and you'll find much more than just plants. A multi-billion-pound industry has grown up around our love affair with gardening, anticipating our every need from patio furniture to a ball of twine.

There are lots of good things about this thriving industry. It gives work to hundreds of thousands of people, contributes millions to the economy, and encourages everyone to do more gardening. That said, there is a heavy carbon cost to pay for the drive to sell gardeners more stuff.

The environmental impact is worst when you're spending your money on well-marketed but poorly designed gadgets you'll use once then consign to the back of the shed. Cheap equipment is often short-lived, too – brightly coloured plastic tools that break easily, or polythene "pop-up" mini-greenhouses that shred in winter gales, end up as landfill fodder within a year.

So, if you're going to buy new, do it responsibly – buy only what you need, avoid environmentally damaging plastic (goods and packaging), and get the best quality you can afford so you can enjoy using it for many years to come.

PRE-LOVED GEAR

Buying any new equipment is a high-carbon option, though, as it's often linear consumption – after the item is made, one owner uses it and then throws it out. That means the energy and resources the item has consumed last a relatively short time.

There is a low-carbon alternative, and it's a lot more fun. Shopping for second-hand gardening gear is like hunting for treasure – you never know quite what you'll find, but chances are it'll be wonderful. You're still consuming, but it's circular – reusing,

ON AVERAGE, WE SPEND TWICE AS MUCH ON **GARDENING GOODS** AS PLANTS. IN **2017** UK GARDENERS SPENT **£1.3 BILLION**; IN THE US IT WAS **$47.8 BILLION**.

"Shopping for second-hand gardening gear is like hunting for treasure."

repairing, and repurposing goods to keep them in use as long as possible.

It adds a whole new dimension to your gardening life. Pick up a pre-loved garden tool and you hold history in your hand; buy an old bench, table, or terracotta pot and it settles into your garden as if it's always been there. Even if you're just looking for day-to-day gear you'll often pay a fraction of the price for a second-hand version.

FINDING SECOND-HAND GARDENING EQUIPMENT

- **Online auction sites and marketplaces:** Ebay, Gumtree, Facebook Marketplace, and Preloved offer rich pickings for second-hand items for the garden; you can source very locally, too.
- **Antique and vintage dealers:** Several specialist dealers offer vintage gardening items, some with online shops. You'll pay more, but you'll find some real gems.
- **Reclamation yards and second-hand markets:** Architectural salvage yards and second-harnd markets are happy hunting grounds for all sorts of vintage features, as well as agricultural equipment that you can repurpose.
- **Auctions:** Look for "general sales" at your local auction rooms and you'll find dirt-cheap items, mainly from house clearances, including lots of gardening accessories.
- **Garden machinery repair shops:** Your local lawnmower service centre can often offer refurbished machines, and sometimes other equipment as well.

SUPPORT YOUR LOCAL ARTISANS

Another way to make your garden shopping a pleasure, while also guaranteeing any garden paraphernalia is sustainable and as low carbon as possible, is to find independent artisans and craftspeople who are making gardening kit locally.

Traditional artisan craft skills are in danger of disappearing in an age of cheap mass production. The UK, for instance, has already lost all its cricket ball makers, gold beaters, and lacrosse stick makers. Perhaps more worryingly for gardeners, sieve makers, hurdle makers, and rake makers are also in short supply.

But you'll likely find blacksmiths, green-wood carpenters, coppice workers, stone masons, thatchers, and dry-stone wallers living and working not far away. So, make good use of their extraordinary skills, experience, and creativity – all of them can help you create your dream garden using locally sourced, sustainable materials with a fraction of the carbon footprint of mass-produced items.

SUPPORT YOUR NEIGHBOURHOOD

Sourcing hand-crafted articles from your area also supports your local economy, helps keep your neighbourhood's history alive, and gives your garden a sense of place – the feeling that everything in it belongs to the wider environment in which it sits.

ARTISANAL MARKETPLACES

You can find local craftspeople at county shows and craft fairs. There are also directories available online (both the Heritage Crafts Association and the Crafts Council have searchable databases) and many, including coppice workers and green woodworkers, have their own professional organizations and guilds.

You can also simply keep your eye out for sustainably produced garden sundries made within your own country, rather than running up a carbon cost by importing goods and materials from overseas. One small business in Devon, England, for example, makes garden twine from the wool of rare breed Whiteface Dartmoor sheep. It's far more sustainable than jute twine imported from India, makes good use of a natural product, and it's lovely and soft on your plants, too.

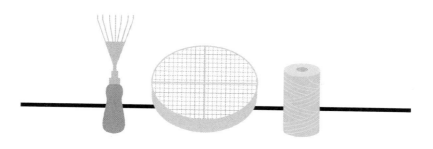

"Sourcing hand-
crafted articles
from your area
supports your
local economy
and gives your
garden a sense
of place."

GROW YOUR OWN GARDEN SUNDRIES

The garden centre with the lowest carbon footprint of all is the one just outside your back door. Your garden can supply you with much of what you need in the way of supports, compost, fertilizers, and labelling, as well as some decorative garden structures.

We've seen how you can produce fertilizers at home (p.106), save your own seed (p.76), and make your own potting compost (p.126) – all items you can cross off your garden centre shopping list for good. If you include useful plants such as hazel (*Corylus avellana*), willow (*Salix*), dogwood (*Cornus*), and New Zealand flax (*Phormium tenax*) in the garden as well, they'll not only look good but keep your shed well stocked, too.

A **COPPICED HAZEL** CAN LIVE FOR UP TO **500 YEARS** – ALMOST **10 TIMES** AS LONG AS A TREE LEFT TO GROW NATURALLY.

GARDEN SUNDRIES TO GROW AT HOME

- **Beanpoles:** Grow coppiced hazel (see p.32) and you get ramrod-straight stems that make excellent plant supports. Harvest three-year-old stems in late winter once they're 3–5cm (1¼–2in) in diameter. Trim off the twiggy tops (but save them for supporting peas) to make poles about 2.5m (8ft) tall.
- **Plant labels:** Cut a hazel twig about 10cm (4in) long and 2cm (¾in) in diameter, and whittle the bark off one end with a sharp knife to make a flat surface. Wooden labels absorb water, smudging your writing, so paint with water-based varnish to seal it or tie the label to the plant instead of sticking it into the compost.
- **Garden twine:** New Zealand flax makes a handsome ornamental garden perennial, but it also has tough, fibrous leaves that provide a never-ending supply of garden twine. Cut a leaf away at the base, split it lengthways, and peel away long, stringy fibres from the edge.
- **Fencing and trellis:** Hazel, willow, and dogwood make wonderful fences, arches, and trellises. Weave young willow or dogwood stems between hazel uprights, or make a simple trellis by lashing together straight hazel stems with flax string to make a grid.
- **Greenhouse shading:** Plant runner beans along the sunny side of a greenhouse and train up strings tied to the roof to provide shade in summer. When you need more light in autumn, simply remove them.

GROW YOUR OWN
SWEET PEA SUPPORT

*Weave together sturdy hazel uprights with young, whippy willow
or dogwood stems to make a beautiful, long-lasting support for
sweet peas, annual climbers, or climbing beans.*

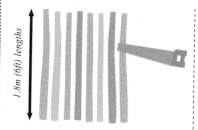

1.8m (6ft) lengths

1 *Cut six to eight straight hazel
stems, 2–3cm (1–1½in) diameter, and
trim to the same length – 1.8m (6ft)
works well.*

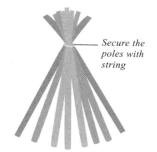

*Secure the
poles with
string*

2 *Push them firmly into the ground
to make a circle, with each pole
crossing at the top. Lash together
firmly with string.*

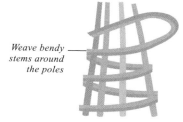

*Weave bendy
stems around
the poles*

3 *Weave brightly coloured stems of
young willow or dogwood stems in
and out between the uprights in wide
bands.* Cornus alba *'Sibirica'
produces brilliant scarlet stems, while
those of* Salix alba var. vitellina *are
flame coloured.*

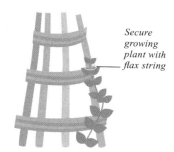

*Secure
growing
plant with
flax string*

4 *Plant one or two young plants at
the foot of each upright, water in, and
tie them in with flax string to guide
them into the support as they grow.*

DEALING WITH GARDEN WASTE

*In the low-carbon garden, there is no such thing
as garden waste – only an array of useful material ready
to lock more carbon back into your garden soil.*

DON'T LET IT GO TO WASTE

There's no such thing as garden waste. This might sound implausible if you're pushing a wheelbarrow of weeds back from the veg patch, but when you view your "waste" as a harvest of useful materials, you start to look forward to getting your hands on more of it.

Traditional ways of getting rid of garden waste all have environmental drawbacks. Garden bonfires give off black carbon, or soot – a major cause of air pollution and climate change. Taking waste to the local tip involves a fossil-fuel-powered car journey, plus transportation to and from the council's recycling depot. If your local council collects green waste from the kerbside, there are still transport costs to factor in – and many councils charge you as well.

Disposing of garden waste on site is carbon-positive, as all the carbon it contains is locked away back into your garden soil. It's also good garden practice. It's all too easy to spread diseases and invasive plants when you send green waste out of the garden. One study found that garden waste in Germany was responsible for almost a third of the notoriously difficult to eradicate Japanese knotweed and nearly 20 per cent of giant hogweed. Besides, reusing waste is so useful. You can recycle it all, from roots to branches, without a shred of it leaving your premises.

WHERE TO REUSE WASTE

- **Leaves and soft stems:** More than 90 per cent of what we throw out is "wet" greens – nitrogen-rich spent flower heads, grass clippings, and weeds. All of it makes wonderful compost. Young annual weeds and the sheared-off top growth of perennial weeds can go in, too.
- **Woody stems:** Use secateurs to cut up spent crops and perennials, plus twiggy prunings less than 2.5cm (1in) thick and compost them,

ACCORDING TO A **STUDY** BY THE UNIVERSITY OF GEORGIA IN THE US, MORE THAN **72 PER CENT** OF **LANDFILL MATERIALS** COULD BE **DIVERTED** THROUGH **COMPOSTING**.

> *"Traditional ways of getting rid of garden waste all have environmental drawbacks."*

spread them over borders as a mulch, or use them to make paths. Feed thicker branches into a living hedge, known as a fedge (see p.52), or build log piles (see p.176).

- **Dead trees:** Stumps and trunks hold masses of carbon, and allowing them to decompose naturally locks it out of the atmosphere for decades, as well as providing five-star habitats for wildlife. Only remove stumps where you identify a honey fungus infection, to limit its spread.

- **Perennial weed roots and seeding weeds:** The roots of perennial weeds, such as bindweed and couch grass, resprout if you add them to the compost alive. Seeds from dock and annual weeds such as bittercress germinate after composting, too. Pile them all into a bucket, cover with water, and leave to rot. After about a month, they'll be dead and safe to compost.

- **Diseased material:** Most common waterborne, foliar fungal diseases, including mildew, rust, and late

Leave stumps in situ.

blight, don't survive after living tissue has decomposed, so it's safe to compost. Foliar diseases, such as rose black spot, are more persistent – bury material at least 30cm (1ft) deep. Only woody material and plants infected by soil-borne diseases such as clubroot must be disposed of off-site – do take these to your council's recycling depot.

MAKE COMPOST

Garden "waste" quickly becomes an asset once it's turned into home-made compost – it's the gardener's secret ingredient for turning any soil into rich, crumbly brown loam your plants will love.

Astonishingly, a third of what we throw in the bin is organic material we could have composted. Instead, it goes to landfill where it's broken down anaerobically (without oxygen) by microbes that release methane. Methane is a greenhouse gas more than 80 times more powerful than carbon dioxide in the first two decades after it's released.

SOIL MULCHED WITH **COMPOST** IS MORE EFFICIENT AT **HOLDING RAINWATER** AND **HOLDS** ON TO **NUTRIENTS BETTER**.

A compost heap uses a different set of microbes. It uses oxygen to digest the carbon stored in your garden waste, turning it into dark-brown organic matter. These microbes do give off some carbon dioxide, but that's easily offset by the powerful boost compost gives to your garden's ability to hold on to carbon.

Soil mulched with compost has a more active biology, and plants grow larger, absorbing more carbon. It is also more efficient at holding rainwater and holding on to nutrients better, so you won't need to water as often or use as much fertilizer.

WHAT TO COMPOST

Any raw vegetable matter makes good compost. Don't add meat, fish, poultry, or dairy products to the compost bin, as they can smell and attract vermin. The same applies to fats and oils. Some of this can go into a wormery instead (see pp.108-109). You can also add manure and animal bedding – but not cat or dog faeces as they harbour germs and parasites. Shredded paper and cardboard are fine, but glossy or coated paper may contain harmful dyes and chemicals.

The bulk of your compost, though, will be made up of garden waste. Cut woody material up small, and make the most of weeds: non-flowering annual weeds plus top growth from nettles, bindweed, and ground elder can all go in. You can even use the roots of perennial weeds if you drown them (see p.173) or dry them first.

Many diseased plants are also safe to compost as the more common fungal diseases that affect foliage, including blight, rust and mildew, don't survive long once the leaves break down.

HOW TO MAKE COMPOST

Build yourself a compost bin from second-hand untreated wood, nailed to wooden uprights with a 1cm (½in) gap between each layer; or screw together four pallets to make a box.

1 *Gradually fill your bin, aiming for a 50:50 mix of fast-rotting "greens" (lawn clippings, leafy growth, vegetable peelings) and slower-rotting "browns" (woody plant material, straw, shredded paper).*

2 *Once the bin is full, wet everything thoroughly with water if the mix is dry, then cover with cardboard to stop weed seeds getting in.*

3 *Turn your compost once or twice to get those microbes multiplying. Fork everything out of the bin, then back in again, treading it down as you go and watering again if necessary.*

4 *Your compost shouldn't smell bad. If it does, or if it turns slimy, it's decomposing anaerobically and emitting methane. If this happens, turn it and mix in extra "browns" to get it back on track again.*

5 *After six months to a year (depending on how often you turned it), you should have a lovely dark-brown, crumbly mix – you shouldn't be able to see individual ingredients any more. Your compost is now ready to use on the garden.*

MAKE A LOG PILE

Rather than disposing of bulky cuttings, stack them in a corner of your garden and create a home for a community of wildlife. Log piles provide a valuable shelter for small mammals and insects, as well as a habitat for mosses, lichens, and fungi.

Every gardener enjoys a good pruning job. There's nothing so satisfying as tackling an overgrown lilac in summer or shaping young apple trees in the depths of December. But all that pruning does generate a lot of woody waste.

DEAD WOOD IN **FORESTS** IS AT AN ALL-TIME **LOW** – FORESTS IN **EUROPE** HAVE LESS THAN **5 PER CENT** OF DEAD WOOD **EXPECTED** IN **NATURAL** CONDITIONS.

For many people, this would be the cue to load up the car and head to the local tip. But if you keep your woody prunings at home, you can lock the carbon they contain back into your garden soil, while doing your bit to revive one of the world's most threatened habitats.

Dead wood has been disappearing from our landscapes as gardeners and park keepers tidy it away, convinced it looks scruffy or might harbour disease. So, you won't often see fallen branches or toppled trees left where they are.

LOG-PILE RESIDENTS

Hundreds of species of creatures and micro-organisms depend on dying and dead wood for survival. This rich habitat evolves over decades, with a different army of organisms colonizing the wood at each stage of decay.

As the wood dies bracket fungi break down the outer layers of lignin (bark). Then, beetles arrive, including rarities like the magnificent 5cm-(2in)-long stag beetle. They're followed by their predators. Frogs, toads, and slow worms love wood piles for their dark, cool shelter and abundant food. If possible, leave a dead or dying tree standing (as long as it's safe), so that birds such as woodpeckers, treecreepers, and owls can move in. All will provide you with free pest control.

If every one of the UK's 24 million gardens (and 300,000 allotments) had at least one log pile, it might turn the tide for the tiny but essential creatures that call them home. You'll also add carbon-rich biomass back into the soil as the wood decays, so your garden begins to actively absorb carbon and combat climate change. And you'll save yourself an awful lot of time traipsing to the tip.

HOW TO CONSTRUCT A LOG PILE

Choose a spot for your log pile away from the house and in dappled shade – full sun dries log piles out, while deep shade is too cool for many insects.

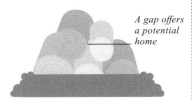

A gap offers a potential home

3 *Stack more wood on top to make a pyramid. Don't worry if it's not too tightly stacked – gaps encourage more creatures to crawl in.*

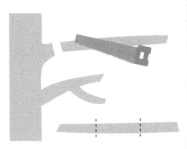

1 *Cut your woody prunings to about 1m (3ft) long. Shorter prunings are fine as they can be tucked in between larger pieces.*

4 *Don't build your pile too high, as the top will dry out. Lots of smaller log piles are better than one big one.*

2 *Use your largest diameter wood for the lowest layer and bury most of it in the ground to keep it damp.*

5 *Add soil here and there as you go. Plant shade-loving species such as ferns and primroses into the sides.*

USE CLIPPINGS AS MULCH

Mulch has many benefits for your garden – it helps to conserve soil moisture, suppress weeds, and protect plants against extreme weather. But there's no need to dash to the garden centre to buy the stuff – save money and help the environment by recycling garden waste to make your own mulch.

A more direct way to return your garden waste to your soil is just to drop it on the ground. You can do this without processing it at all, but it looks a bit messy, so it's better to cut up your clippings into short pieces as you prune. If you have large quantities, it's more efficient to shred them, ideally with a shredder powered by electricity from renewable sources.

The resulting dense, woody mulch is immensely useful around the garden. A generous layer 8–13cm (3–5in) thick acts like a duvet over

---

A **WELL-MULCHED** GARDEN CAN YIELD **50 PER CENT** MORE **VEGETABLES** THAN AN **UNMULCHED** GARDEN.

---

the soil, protecting it from the winter weather, as well as holding in moisture and keeping weeds at bay. And, of course, all the time your

shredded prunings are adding the carbon they hold back into the soil. Ideally, stack your prunings first for three or four months so any toxic compounds they might contain can break down. Avoid digging in woody mulches, as they then decay under the soil and might "rob" it of nitrogen, causing temporary nutrient deficiencies right where your plants are growing.

MAKE A MULCH PATHWAY

Home-made woody mulch is a wonderful cost-free material for garden paths. Use fresh clippings to make soft, natural paths. Level the ground first, removing weeds, and scalping any turf as low as you can get it. Cover the ground with thick sheets of cardboard to make sure nothing grows back through.

Edge your path with planks, reclaimed bricks, or stone, then pile your woody prunings in a layer at least 10cm (4in) thick. Top this up continuously throughout the season.

"Woody mulches are particularly long lasting, as they're slow to decay."

TAKING IT FURTHER

Your garden is surprisingly rich in resources that can help reduce your carbon emissions not only when you're gardening but in all areas of your life.

GROW USEFUL PLANTS

Home-made is low carbon. Whenever you make something yourself, using ingredients grown in your garden, you're bypassing all the emissions involved in manufacture, packaging, transporting, and selling its commercial alternative.

Most of the carbon emissions you're responsible for happen outside the garden fence. As you go about your day, heating (or air conditioning) your house, going shopping, travelling to and from work, and generally living your life, you are constantly adding greenhouse gases to the atmosphere.

STUDIES HAVE SHOWN THAT **SAGE** MAY **BENEFIT COGNITIVE** HEALTH AND POTENTIALLY **HEART** AND **COLON** HEALTH.

As we transfer to more sustainable energy sources, emissions are steadily falling, but still not fast enough. We all have to be more aware of the demands we make on resources. And here your garden can step in to help.

Obviously you can't make everything you use in daily life, but every little helps. Besides, crafting your own soaps, or brewing home-grown herbal tea can be so enjoyable that once you start you'll want to do more, until self-reliance becomes a firm habit and the foundation of a truly low-carbon life.

GROW YOUR OWN TEA

Classic "black" tea comes from the evergreen shrub *Camellia sinensis*. It's easy to grow as long as you have acidic soil (if not, grow in containers of peat-free ericaceous compost).

Herbal teas are easier: just pick a 10cm (4in)-long sprig and steep for five minutes in boiling water. The following are good herbal teas.

- **Peppermint** (*Mentha* x *piperita*): Mint tea is a pick-me-up and also helps digestion, so drink it after eating to settle your stomach.
- **Chamomile** (*Matricaria recutita* or *Chamaemelum nobile*): A teaspoon of dried chamomile flowers brews a calming, relaxing tea.

> *"Self-reliance becomes a firm habit
> and the foundation of a truly
> low-carbon life."*

- **Rosemary** (*Salvia rosmarinus*): A cup of rosemary tea acts as a natural stimulant and helps improve your memory, too.

GROW YOUR OWN SKINCARE

Bypass the sometimes harsh chemical ingredients in commercial beauty treatments and grow your own. It needn't stop at moisturizer – you can make shampoo, soap, and bath oil.

- **Soapwort** (*Saponaria officinalis*): A perennial with pink flowers, its roots and leaves are rich in saponin, which makes excellent soap and shampoo.
- **Lavender** (*Lavandula angustifolia*): Hang a bag of dried lavender flowers under the hot tap for a calming, relaxing, and beautifully scented bedtime bath.
- **Pot marigold** (*Calendula officinalis*): Cheery orange or yellow calendula flowers are naturally healing and make soothing, gentle creams for sensitive skin.

GROW YOUR OWN MEDICINE CABINET

Home-made medicines have their limits – if there's something seriously wrong, you should always see a doctor. But for many more everyday maladies, you can just pick your cure from the natural world.

- **Aloe vera:** The sap of this succulent houseplant is a natural healing gel – snap a leaf in half and rub the gooey, soothing juice onto minor burns, including sunburn.
- **Sage** (*Salvia officinalis*): Make a tea from hardy, evergreen sage and gargle with it three times a day to soothe a sore throat.
- **Rosehips:** Traditional rosehip syrup is packed with vitamin C to boost the immune system; a spoonful each morning acts like a home-made shot of multivitamins. Any rose with hips will do but *Rosa rugosa* has the biggest and fattest.

GROW A SQUARE-FOOT CUT-FLOWER PATCH

Using your garden to grow vegetables (see pp.70–77) is one of the most effective ways you can help tackle climate change. And if you often pick up a bunch of flowers at the supermarket, your garden could help to reduce your carbon footprint even further.

Out-of-season cut flowers are among the most carbon-heavy items you can buy, pound for pound. Lilies imported from the Netherlands carry the highest footprint, at nearly 3.5kg (7¾lb) of carbon a stem, as they're grown in heated greenhouses. Kenyan-grown roses and gypsophila, transported to you by air, come in at about 2.4kg (5⅓lb) a stem.

ROSES GROWN FOR **VALENTINE'S DAY** IN THE US PRODUCE ABOUT **9,000 TONNES (9,920 TONS)** OF **CARBON** EMISSIONS.

CUT FLOWERS THE LOW-CARBON WAY

A bunch of locally grown, seasonal, cut flowers is much better, at about a tenth the emissions. But best of all are the flowers you grow yourself. If you use the techniques outlined in this book, they carry just a fraction of the emissions of bought flowers.

- **Pick your existing flowers:** The no-fuss way to fill your home with flowers is simply to head into the garden and pick a posy of whatever's in bloom. It's a lovely way of bringing the garden indoors.
- **Make a perennial cut-flower patch:** Pick as much as you want from a dedicated flower patch filled with long-lived flowering shrubs and perennials, chosen to lock in carbon as well as to look good in a vase. Mix roses, hydrangeas, and lilacs with long-stemmed perennials such as aquilegia, phlox, penstemons, and perennial sunflowers for season-long colour.
- **Grow hardy annuals:** If you prefer a wilder, lighter look, fill your flower patch with annuals you can sow directly into the soil. Then you can save the seed to keep it carbon neutral – they may not come back the same each year, but it's fun to see what turns up! Good choices include sweet peas, cosmea (*Cosmos bipinnatus*), and poppies.
- **Grow roses:** If you like roses, the following are all great choices for cut flowers. *Rosa xanthina* 'Canary Bird', 'Cardinal de Richelieu', 'Tuscany Superb', 'Charles de Mills', *Rosa gallica* 'Versicolor' (also known as *Rosa mundi*), 'New Dawn', and 'Souvenir du Docteur Jamain'.

PLANT A SQUARE-FOOT CUTTING GARDEN

Square-foot gardens are usually for growing vegetables, but they work well for cut flowers, too, letting you grow a huge variety of blooms in a tiny space.

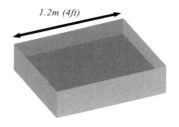

1.2m (4ft)

1 *Build a raised bed at least 1.2m (4ft) square and fill with a 50:50 mix of topsoil dug from the garden and home-made compost.*

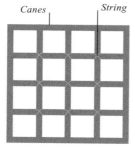

Canes   String

2 *Using canes and string, divide the bed into equal sections no smaller than 30 x 30cm (1sq ft); a larger area will provide more flowers.*

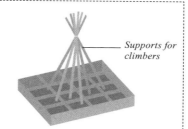

Supports for climbers

3 *Put up supports for climbers such as sweet peas – sink 1.8m (6ft) hazel rods into four squares, then tie together into a tripod. Sow two seeds to each upright.*

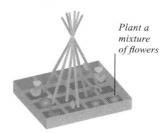

Plant a mixture of flowers

4 *Fill each square with a different flower. A 30 x 30cm (1sq ft) square accommodates four annuals, or one larger perennial.*

Nourish the plants with comfrey tea

5 *Feed your flowers weekly with comfrey tea (see p.106) and resow picked-out squares straight away to keep the flowers coming.*

CLOTHE YOUR HOUSE IN GREENERY

When you're feeling a bit chilly, the first thing you reach for is a toasty wool blanket or a nice warm coat to hold in the heat. So you'll not be surprised that it's the same for your home.

An awful lot of the energy (and, therefore, greenhouse gas emissions) we use in heating our homes in winter simply escapes, mainly through the walls (36 per cent) and the roof (20 per cent). Green roofs and walls form a barrier, stopping cold air and wind from entering your home through cracks, and trapping a layer of warmer air behind their leaves. Green roofs also capture and hold on to rainwater rather than letting it run off.

COOLING EFFECT

Houses, made of warmth-absorbing brick and dark-coloured, tiled roof surfaces, also act as giant heat sinks in summer, forcing us to switch on the fans just to stay comfortable.

Throw a thick coat of plants over your home and you hold the warmth in during winter and keep the heat out in summer. Plus, you double down on reducing greenhouse gas emissions – you use less energy to keep yourself comfy, and your plants absorb carbon from the air as they grow.

GREEN LAYERS

Green roofs and walls are a wonderful way to insulate and shade your buildings.

Green walls *High-tech green wall installations involve complicated engineering and require artificial irrigation and feeding, using up lots of resources to manufacture, install, and maintain. So, create your own living wall. As long as your masonry is sound, you can train climbing plants up the wall of your house on sturdy wires for a low-tech and much more sustainable version that's equally energy-saving.*

IN THE WINTER

Thermal mass *Thick, dense greenery alongside house walls, and soil as well as plants in a green roof, add thermal mass, absorbing warmth from your house and holding it there so it doesn't escape.*

The blanket effect *Vegetation provides a physical protection from the chilling effects of cold wind. It also fends off the worst of the winter rains. Contrary to popular opinion, research has shown climbing plants don't make walls damp – they protect them by keeping humidity levels constant.*

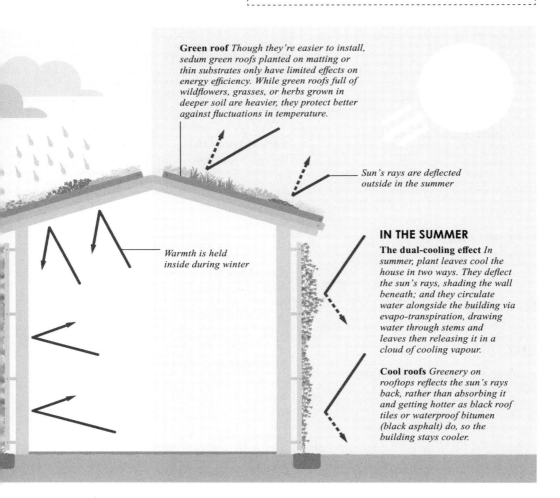

Green roof *Though they're easier to install, sedum green roofs planted on matting or thin substrates only have limited effects on energy efficiency. While green roofs full of wildflowers, grasses, or herbs grown in deeper soil are heavier, they protect better against fluctuations in temperature.*

Sun's rays are deflected outside in the summer

Warmth is held inside during winter

IN THE SUMMER

The dual-cooling effect *In summer, plant leaves cool the house in two ways. They deflect the sun's rays, shading the wall beneath; and they circulate water alongside the building via evapo-transpiration, drawing water through stems and leaves then releasing it in a cloud of cooling vapour.*

Cool roofs *Greenery on rooftops reflects the sun's rays back, rather than absorbing it and getting hotter as black roof tiles or waterproof bitumen (black asphalt) do, so the building stays cooler.*

FURTHER RESOURCES

CLIMATE CHANGE, CARBON EMISSIONS, AND GARDENING

Berners-Lee, M. (2010). *How Bad are Bananas? The Carbon Cost of Absolutely Everything.* Green Profile.
Blanusa, T., and Page, A. (2011). *RHS Gardening Matters: Urban Gardens.* Royal Horticultural Society, UK.
Morgan, S., and Stoddart, K. (2019). *The Climate Change Garden.* Green Rocket.
Webster, E., Cameron, R.W.F., and Culham, A. (2017). *RHS Gardening in a Changing Climate.* Royal Horticultural Society, UK.
RHS online publications:
rhs.org.uk/science/pdf/RHS-Gardening-in-a-Changing-Climate-Report.pdf
rhs.org.uk/science/pdf/climate-and-sustainability/urban-greening/gardening-matters-urban-greening.pdf

SUSTAINABLE GARDEN DESIGN

Dunnett, N. (2019). *Naturalistic Planting Design The Essential Guide: How to Design High-Impact, Low-Input Gardens.* Filbert Press.
Weaner, L. (2016). *Garden Revolution: How Our Landscapes can be a Source of Environmental Change.* Timber Press.

TAPESTRY LAWNS

Smith, L. (2019). *Tapestry Lawns: Freed from Grass and Full of Flowers.* CRC Press.
grassfreelawns.co.uk

REWILDING

Tree, I. (2019). *Wilding: the Return of Nature to a British Farm.* Picador.
plantlife.org.uk/everyflowercounts

BIODIVERSITY AND GARDENING FOR WILDLIFE

Goulson, D. (2020). *The Garden Jungle, or Gardening to Save the Planet.* Vintage.
Dogwood Days blog list of peat-free nurseries:
dogwooddays.net/2020/04/30/updated-peat-free-nurseries-list
RHS pesticide-free nurseries:
rhs.org.uk/advice/profile?PID=960

NO-DIG GARDENING

Dowding, C. and Hafferty, S. (2017). *No Dig Organic Home and Garden.* Permanent Publications.
Fukuoka, M. (2009). *The One-Straw Revolution.* NYRB Classics.
charlesdowding.co.uk

PERMACULTURE AND FOREST GARDENING

Crawford, M. (2012). *How to Grow Perennial Vegetables: Low-maintenance, Low-impact Vegetable Gardening.* Green Books.
Holzer, S. (2010). *Sepp Holzer's Permaculture.* Permanent Publications.
Shein, C. and Thompson, J. (2013). *The Vegetable Gardener's Guide to Permaculture: Creating an Edible Ecosystem.* Timber Press.
backyardlarder.com

RAIN GARDENING

Dunnett, N. and Clayden, A. (2007). *Managing Water Sustainably in the Garden and Landscape.* Timber Press.

WEEDS

Mabey, R. (2012). *Weeds: The Story of Outlaw Plants.* Profile Books.
Wallington, J. (2019). *Wild about Weeds.* Laurence King Publishing.

GREEN ROOFS

Gedge, D. and Little, J. (2008). *The DIY Guide to Green and Living Roofs.*
greenrooftraining.com
grassroofcompany.co.uk
livingroofs.org

COMPOST AND SOIL SCIENCE

Lowenfels, J. (2010). *Teaming with Microbes: The Organic Gardener's Guide to the Soil Food Web.* Timber Press.

SEED SAVING

McVicar, J. (2012). *RHS Seeds.* Kyle Cathie.
realseeds.co.uk

ACKNOWLEDGMENTS

AUTHOR'S ACKNOWLEDGMENTS

My first thanks go to Chris Young, Editor of *The Garden* and Consultant Gardening Publisher for DK, for listening patiently to my rants about plastic in the garden and believing in me enough to commission a year-long series of articles on the subject. It gave me the space to say all I had been bottling up for years about the sustainability (or not) of gardening, and crystallized for me the thought that this was a topic too important to relegate to my experimental tinkerings in the garden.

Thanks also go to Mary-Clare Jerram at DK for picking up the idea and running with it, and the rest of the wonderful and smoothly efficient team at DK – especially Ruth O'Rourke, Joy Evatt, Nikki Sims, and Dawn Titmus for their patience and reassuringly sure-footed editing.

I'd also like to thank the many people at the RHS who have given generously of their time and insights to help me get to grips with a complex scientific field, particularly Dr Tijana Blanusa, Helen Bostock, and Guy Barter, whose constructive comments have added greatly to the depth of this book.

I am also very grateful to the many gardeners whose pioneering work has informed these pages throughout and hugely influences everything I do in my garden: Charles Dowding, Professor Nigel Dunnett, Professor Dave Goulson, Martin Crawford, John Little, Dusty Gedge, Lionel Smith, and Isabella Tree have all been inspirational.

Finally, I'd like to thank my two lovely daughters, Ellie and Ruby, who somehow remained stalwart and supportive through months of mild neglect, brought me coffee, and generally kept me cheerful.

Thanks to the RHS Science team for their research, enthusiasm, and input into the reading of this book.

PUBLISHER'S ACKNOWLEDGMENTS

Dorling Kindersley would like to thank Jane Simmonds for proofreading and Vanessa Bird for indexing.

ABOUT THE AUTHOR

Sally Nex is a professional gardener and garden writer, and also teaches and gives lectures about food growing and sustainability. After a long career as a BBC News journalist in central London, she gave up the day job to follow her love of horticulture – combining journalism with gardening and a passion for the environment. She now writes regularly for the RHS magazine *The Garden*, as well as BBC *Gardeners' World* magazine, the *Telegraph Gardening*, and *Grow Your Own* magazine, with a special interest in growing sustainably and organically.

INDEX

Senior Editors Dawn Titmus, Nikki Sims
Editor Joy Evatt
Designer Amy Cox
Project Art Editors Jessica Tapolcai, Harriet Yeomans
Illustrators Amy Cox, Keith Hagan
Senior Production Editor Tony Phipps
Production Controller Rebecca Parton
Jacket Designer Amy Cox
Jacket Coordinator Lucy Philpott
Managing Editor Ruth O'Rourke
Managing Art Editor Christine Keilty
Art Director Maxine Pedliham
Publishing Directors Katie Cowan, Mary-Clare Jerram
Consultant Gardening Publisher Chris Young

ROYAL HORTICULTURAL SOCIETY
RHS Publisher Rae Spencer-Jones

First published in Great Britain in 2021 by
Dorling Kindersley Limited
One Embassy Gardens, 8 Viaduct Gardens,
London SW11 7BW

The authorized representative in the EEA is
Dorling Kindersley Verlag GmbH. Arnulfstr. 124,
80636 Munich, Germany

A CIP catalogue record for this book
is available from the British Library.
ISBN: 978-0-2414-7297-2

Printed in Latvia

For the curious
www.dk.com

MIX
Paper from
responsible sources
FSC™ C018179

This book was made with Forest Stewardship Council™ certified paper
– one small step in DK's commitment to a sustainable future.
For more information go to
www.dk.com/our-green-pledge